STRUCTURAL STEEL DESIGN
LRFD FUNDAMENTALS

J. C. Smith
North Carolina State University

D0148887

John Wiley & Sons
New York Chichester Brisbane Toronto Singapore

LaserWriter is a registered trademark of Apple Computer, Inc.
MacDraw is a registered trademark of CLARIS Corporation.
Macintosh is a trademark of Apple Computer, Inc.
MathWriter is a trademark of Cooke Publications.
Microsoft Word is a trademark of Microsoft Corporation.

TA
684
.S585
1988

Copyright © 1988 by John Wiley & Sons, Inc.

All rights reserved.

Reproduction or translation of any part of
this work beyond that permitted by Sections
107 and 108 of the 1976 United States Copyright
Act without permission of the copyright owner
is unlawful. Requests for permission or further
information should be addressed to the
Permissions Department, John Wiley & Sons.

ISBN 0-471-62141-2

Printed in the United States of America.

10 9 8 7 6 5 4 3 2 1

DISCARDED
WIDENER UNIVERSITY
WIDENER UNIVERSITY
WOLFGRAM
LIBRARY
CHESTER, PA.

To My Family
Lois, Jonathan, Olivia, Keith, and Nancy

PREFACE

This book was written to serve as the undergraduate level textbook for the first structural steel design course in Civil Engineering when the AISC Load and Resistance Factor Design(LRFD) Specifications are used. Some of the material in Chapters 1 and 6 may be more appropriate for the second structural steel design course. Also, this book is intended to serve as a brief introduction to the LRFD Approach for structural design practitioners.

When the LRFD Manual became available, I began to use it in two graduate level courses. One course deals only with plastic analysis and design of continuous beams and plane frames. The other course deals with buckling of frames, second order effects, elastic analysis, and computer aided design of multistory buildings. After these two favorable experiences with the LRFD Manual, I decided to begin using the LRFD approach in the first undergraduate steel design course. Since reliable sources revealed that the first LRFD textbook probably would not be available until about the fall semester of 1989, I began to write a camera ready textbook which would contain only the material I planned to teach in the first undergraduate steel design course. I had always begun with tension members in the ASD approach for the first undergraduate steel design course, but I had not dealt with the LRFD approach for tension members in the two graduate level courses. Consequently, the first chapter I wrote was on the behavior and design of tension members by the LRFD approach. There is not time enough in the first undergraduate course to discuss all types of connections, so I decided not to include a chapter on connections. However, the behavior and design of tension members is dependent on the type of member end connections. Therefore, I chose to include a very brief introduction to fillet welded and bearing-type bolted connections for tension members at appropriate locations in Chapter 3. Thus, I treated member end connections for tension members as an integral part of the behavior and design process of tension members whereas in most texts connections are discussed in a separate chapter.

Although the LRFD Specifications permit either an elastic factored load analysis or plastic analysis, my decision to use only the elastic factored load analysis was almost automatic. We have a second structural design course at the undergraduate level in which the students are required to design a six story office building as an unbraced frame using steel members and to design the same building using reinforced concrete members. I have written a computer program to perform a second order plastic analysis of a plane frame and use it in the graduate level courses. However, the ACI Code permits only an elastic factored load analysis. I did not want to require my students to learn how to use two different computer analysis packages, so I decided to use only the elastic factored load analysis approach in the second undergraduate structural design course. When I made this decision, I also decided to include some textual material which would give structural design practitioners and the students in our second undergraduate steel design course a brief but realistic introduction to analysis and design of unbraced frames in the LRFD approach. Consequently, I included the last two sections of Chapter 6 and Appendix A.

I do not use the textual material associated with Appendix A in the first undergraduate course. Also, I do not use any of the text example problems in my classroom presentations. Instead, I make up simpler and completely different examples for my classroom presentations. This provides the students with at least two examples which they can study for a thorough understanding of what is being numerically illustrated.

Appendix B was included to serve as review material since the students need a thorough understanding of principal axes for column action and for beam action.

Until the students become familiar with the organization of the LRFD Manual and Specifications, they do not know precisely where to look for the needed information. Therefore, throughout this text I chose to give the applicable page numbers in the LRFD Manual in addition to the applicable LRFD Specification numbers. The applicable page numbers in the LRFD Manual were given to aid the students in quickly finding the needed information.

ACKNOWLEDGEMENTS

I hereby acknowledge that Harper & Row, Publishers, Inc. granted me permission to use the following indicated portions of my textbook entitled *Structural Analysis* [1] copyright 1988:
1. Pages 1 through 14 of *Structural Analysis* [1] except for articles entitled Behavioral Assumptions and Methods of Structural Analysis are reprinted herein as pages 1 to 16, 20, and 21 by permission of Harper & Row, Publishers, Inc.
2. Pages 137 and 138 of *Structural Analysis* [1] are reprinted herein as pages 36, 37, 159, and 160 by permission of Harper & Row, Publishers, Inc.

Also, I appreciate the list of detected errors and detailed comments on chapter 3 given to me by my teaching assistant, Mr. E. T. McMurray, during the first semester this text was being used in an undergraduate steel design class.

When I encountered something that could not be done as described in the Macintosh II user's manual and in the software manuals, I went to see my Mac II guru, John F. Ely, who is also my colleague. He had always been there before me and correctly told me how to do what I was wanting to do. He saved me from numerous hours of frustration and I thank him profusely.

Also, I must thank Cliff Robichaud, engineering editor of John Wiley & Sons, for being confident that I would prepare an acceptable camera ready textbook.

J. C. Smith

October 1988

A Note on the Production of this Text

An Apple Macintosh II personal computer having only 1MB memory and a 40MB hard disk was used in the preparation of this text. The author printed the text on an Apple LaserWriter and Wiley photographed the LaserWriter copy to obtain the camera ready pages. The software applications used were Microsoft Word for word processing, MacDraw I for the graphics, and MathWriter for the more complex equations. The Times 12 point font was chosen for the text, and either 9 or 10 point fonts were chosen for the subscripts and superscripts.

About the Author

J. C. Smith was born in 1933 near Hudson, North Carolina and lived there until he began his studies in the Civil Engineering Department at North Carolina State University where he received his B.C.E. degree in 1955 and entered graduate school. His master's degree studies were interrupted by a two year tour of duty in the U. S. Army where he served as a Civil Engineering Assistant doing planning studies on the blast and fallout effects of nuclear weapons. He received his M.S. degree from North Carolina State University in 1960, became an Instructor for one year, and began his doctoral studies. He received a faculty Ford Foundation Grant for two years to pursue his doctoral studies in Civil Engineering at Purdue University where he received his Ph.D. degree in January 1966. In 1965 he accepted a faculty position at North Carolina State University where he is now an Associate Professor of Civil Engineering. He is the author or coauthor of 20 technical papers and the author of 3 other texts. His areas of expertise are in the analysis and design of steel and reinforced concrete structures with special emphasis on computer applications. He is a registered professional engineer in the state of North Carolina. On numerous occasions he has been asked to serve as a consultant to structural engineering firms and to privately owned companies engaged in construction and in engineering. He is a Fellow of the American Society of Civil Engineers. He has been Secretary-Treasurer of the North Carolina Section of ASCE, Vice-President of the North Carolina Section of ASCE, and the Faculty Advisor of the Student Chapter of ASCE at North Carolina State University.

CONTENTS

Chapter 1

INTRODUCTION

1.1 STRUCTURAL BEHAVIOR, ANALYSIS, AND DESIGN

A *structure* is an assembly of members interconnected by joints. A *member* spans between two joints. The points at which two or more members of a structure are connected are called *joints*. Each support for the structure is a boundary joint which is prevented from moving in certain directions as defined by the structural designer.

Structural behavior is the response of a structure to applied loads and environmental effects (wind, earthquakes, temperature changes, snow, ice, rain, etc.).

Structural analysis is the determination of the reactions, member forces, and deformations of the structure due to applied loads and environmental effects.

Structural design involves:

1. arranging the general layout of the structure to satisfy the owner's requirements (for non-industrial type buildings, an architect usually does this part)
2. preliminary cost studies of alternative structural framing schemes or/and materials of construction
3. preliminary analyses and designs for one or more of the possible alternatives studied in item 2
4. choosing the alternative to be used in the final design
5. performing the final design which involves:
   a) choosing the analytical model to use in the analyses
   b) determination of the loads
   c) performing the analyses using assumed member sizes which were obtained in the preliminary design phase
   d) using the analysis results to determine if the trial member sizes satisfy the design code requirements
   e) resizing the members, if necessary, and repeating items b) thru e) of step 5, if necessary
6. checking the steel fabricator's shop drawings to ensure that the fabricated pieces will fit together properly and behave properly after they are assembled

1

7.   inspecting the structure as construction progresses to ensure that the erected structure conforms to the structural design drawings and specifications

Structural analysis is performed for structural design purposes. In the design process, members must be chosen such that design specifications for deflection, shear, bending moment and axial force are not violated. Design specifications are written in such a manner that separate analyses are needed for *dead loads* (permanent loads), *live loads* (position and/or magnitude vary with time), *snow loads*, and effects due to *wind* and *earthquakes*. Influence lines may be needed for positioning live loads to cause their maximum effect. In addition, the structural designer may need to consider the effects due to fabrication and construction tolerances being exceeded; temperature changes; and differential settlement of supports. Numerical values of E and I must be known to perform continuous beam analyses due to differential settlement of supports, but only relative values of EI are needed to perform analyses due to loads.

Structural engineers deal with the analysis and design of buildings, bridges, conveyor support structures, cranes, dams, offshore oil platforms, pipelines, stadiums, transmission towers, storage tanks, tunnels, pavement slabs for airports and highways, and structural components of airplanes, spacecraft, automobiles, buses, and ships. The same basic principles of analysis are applicable to each of these structures.

The engineer in charge of the structural design must: 1) decide how it is desired for the structure to behave when the structure is subjected to applied loads and environmental effects; and, 2) ensure that the structure is designed to behave that way. Otherwise a designed structure must be studied to determine how it responds to applied loads and environmental effects. These studies may involve making and testing a small scale model of the actual structure to determine the structural behavior (this approach is warranted for a uniquely designed structure -- no one has ever designed one like it before). Full scale tests to collapse are not economically feasible for one of a kind structures. For mass produced structures such as airplanes, automobiles, and multiple unit (repetitive) construction, the optimum design is needed and full scale tests are routinely made to gather valuable data which are used in defining the analytical model employed in computerized solutions.

Analytical models (some analysts prefer to call them mathematical models) are studied to determine which analytical model best predicts the desired behavior of the structure due to applied loads and

2

environmental effects. Determination of the applied loads and the effects due to the environment is a function of the structural behavior, any available experimental data, and the designer's judgment based on experience.

A properly designed structure must have adequate *strength*, *stiffness*, *stability*, and *durability*. The applicable structural design code is used to determine if a structural component has adequate *strength* to resist the forces required of it based on the results obtained from structural analyses. Adequate *stiffness* is required, for example, to prevent excessive deflections and undesirable structural vibrations. There are two types of possible *instability* : 1) a structure may not be adequately configured either externally or internally to resist a completely general set of applied loads; and, 2) a structure may buckle due to excessive compressive axial forces in one or more members. A skateboard, for example, is externally unstable in its length direction. If a very small force is applied to a skateboard in its length direction, the skateboard begins to roll in that direction. Overall internal structural *stability* of determinate frames may be achieved by designing either truss-type bracing schemes or shear walls to resist the applied lateral loads. In the truss-type bracing schemes, members which are required to resist axial compression forces must be adequately designed to prevent buckling, otherwise the integrity of the bracing scheme is destroyed. Indeterminate structural frames do not need shearwalls or truss-type bracing schemes to provide the lateral stability resistance required to resist the applied lateral loads. However, indeterminate frames can become unstable due to sidesway buckling of the structure.

In the coursework that an aspiring structural engineer must master, the traditional approach has been to teach at least one course in structural analysis and to require that course as a prerequisite for the first course in structural member behavior and design. This traditional approach of separately teaching analysis and design is the proper one in the author's opinion, but in this approach the student is not exposed to the true role of a structural engineer unless the student takes a structural design course which deals with the design of an entire structure. In the design of an entire structure, it becomes obvious that structural behavior, analysis, and design are inter-related. The most bothersome thing to the student in the first design of an entire structure using plane frame analyses is the determination of the loads and how they are transferred from floor slab to beams, from beams to girders, from girders to columns, and from columns to supports. Transferral of the loads is dependent on the analytical models which are deemed to best represent the behavior of the structure. Consequently, in the first structural design courses the analytical model and the applied loads are

3

given information, and the focus is on structural behavior and learning how to obtain member sizes that satisfy the design specifications.

1.2 IDEALIZED ANALYTICAL MODELS

Structural analyses are conducted on an analytical model which is an idealization of the actual structure. Engineering judgment must be used in defining the idealized structure such that it represents the actual structural behavior as accurately as is practically possible. Certain assumptions have to be made for practical reasons: idealized material properties are used, estimations of the effects of boundary conditions must be considered, and complex structural details that have little effect on the overall structural behavior can be ignored (or studied later as a localized effect after the overall structural analysis is obtained).

All structures are three dimensional, but in many cases it is possible to analyze the structure as being two dimensional in two mutually perpendicular directions. This text deals only with either truss type or frame type structures. If a structure must be treated as being three dimensional, in this text it is classified as being either a space truss or a space frame.

A *truss* is a structural system of members that are assumed to be pin connected at their ends. Truss members are designed to resist only axial forces and truss joints are designed to simulate a no moment resistance capacity.

A *frame* is a structural system of members that are connected at their ends to joints that are capable of receiving member end moments and capable of transferring member end moments between two or more member ends at a common point.

If all of the members of a structure lie in the same plane, the structure is a two dimensional or planar structure. Examples of planar structures shown in Figure 1.1 are: beams, plane grids, plane frames, and plane trusses. Note that all members of a plane grid lie in the same plane, but all loads are applied perpendicular to that plane. For all of the other planar structures, all applied loads and all members of the structure lie in the same plane. In Figure 1.1 each member is represented by only one straight line between two joints. Each joint is assumed to be a point which has no size. Members have dimensions of depth and width, but a single line is chosen for graphical convenience to represent the member spanning between two joints. Thus, the idealized structure is a line diagram configuration. The length of each line defines the span length of a member and usually each line is the

4

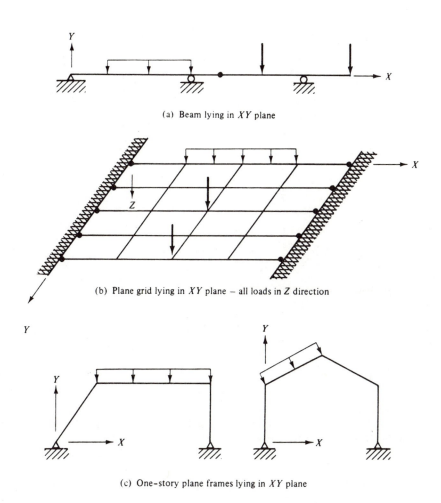

(a) Beam lying in XY plane

(b) Plane grid lying in XY plane – all loads in Z direction

(c) One-story plane frames lying in XY plane

Figure 1.1 Examples of planar structures

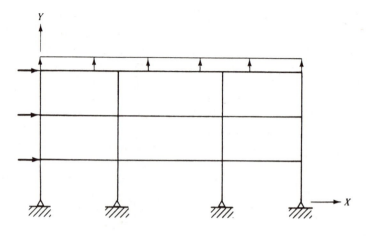

(d) Multistory, multibay, plane frame lying in XY plane

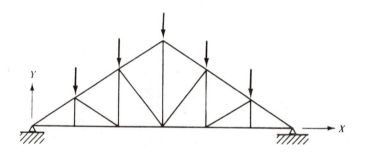

(e) Plane truss lying in XY plane

Figure 1.1 (continued)

trace along the member's length of the intersecting point of the centroidal axes of the member's cross section.

1.3 BOUNDARY CONDITIONS

For simplicity purposes in the following discussion, the structure is assumed to be a plane frame. At one or more points on the structure, the structure must be connected either to a foundation or to another structure. These points are called support joints (or boundary joints, or exterior joints). The manner in which the structure is connected to the foundation and the design of the foundation influence the number and type of restraints provided by the support joints. Since the support joints are on the boundary of a structure and since special conditions can exist at the support joint locations, the term boundary conditions is used for brevity purposes to embody the special conditions that exist at the support joints. The various idealized boundary condition symbols for the line diagram structure are shown in Figure 1.2 and discussed in the following paragraphs.

A *hinge*, Figure 1.2(a), represents that a structural part is pin connected to a foundation which does not allow translational movements in two mutually perpendicular directions. The pin connection is assumed to be frictionless. Therefore, the attached structural part is completely free to rotate with respect to the foundation. Since many of the applied loads on the structure are caused by and act in the direction of gravity, one of the two mutually perpendicular support directions is chosen to be parallel to the gravity direction. In conducting a structural analysis, the analyst guesses that the correct direction of this support force component is either opposite to the direction of the forces caused by gravity or in the same direction as the forces caused by gravity. In Figure 1.2, the reaction components are shown as vectors with a slash on them and the arrows indicate the author's choice for the guessed directions of each vector. It must be noted that the author could have guessed the opposite direction for each vector.

A *roller*, Figure 1.2(b), represents a foundation that permits the attached structural part to rotate freely with respect to the foundation and to translate freely in the direction parallel to the foundation surface, but does not permit any translational movement in any other direction. To avoid any ambiguity for a roller on an inclined surface, Figure 1.2(c), the author prefers to use a different roller symbol than he uses on a horizontal surface. A *link* is defined as being a fictitious, weightless, non-deformable, pinned-ended member that never has any loads applied to it except at the ends of the member. Some analysts prefer to use a link, Figure 1.2(d), instead of a roller to represent the boundary condition described at the beginning of this paragraph.

A *fixed support*, Figure 1.2(e), represents a bed rock type of foundation that does not deform in any manner whatsoever and the structural part is attached to the foundation in such a manner that no relative movements can occur between the foundation and the attached structural part.

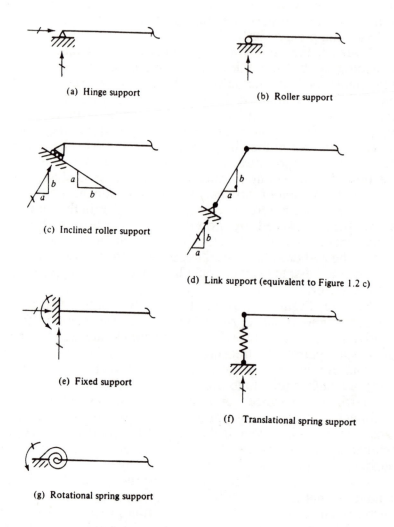

(a) Hinge support

(b) Roller support

(c) Inclined roller support

(d) Link support (equivalent to Figure 1.2 c)

(e) Fixed support

(f) Translational spring support

(g) Rotational spring support

Figure 1.2 Boundary condition symbols and their reaction
components

8

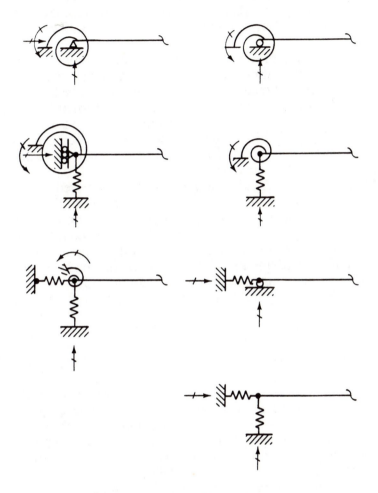

(h) Various possible combinations of Figures 1.2 a through g

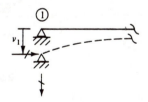

(i) Prescribed Y-direction support displacement

Figure 1.2 (continued)

9

A *translational spring*, Figure 1.2(f), is a link which can deform only along its length direction. This symbol is used to represent either a joint in another structure or a foundation resting on a deformable soil.

A *rotational spring*, Figure 1.2(g), represents a support that provides some rotational restraint for the attached structural part, but does not provide any translational restraint. The support can be either a joint in another structure or a foundation resting on a deformable soil. Generally, as shown in Figure 1.2(h), a rotational spring is used in conjunction with either a hinge, or a roller, or a roller plus a translational spring, or a translational spring, or two mutually perpendicular translational springs.

The soil beneath each individual foundation is compressed by the weight of the structure. Soil conditions beneath all individual foundations are not identical. The weights acting on the foundations are not identical and vary with respect to time. Therefore, nonuniform or differential settlement of the structure occurs at the support joints. Estimated differential settlements of the supports are made by the foundation engineeer and treated as prescribed support movements by the structural engineer. Figure 1.2i shows a *prescribed support movement* .

1.4 INTERIOR JOINTS

For simplicity and generality purposes in the following discussion, the structure is assumed to be a plane frame. On a line diagram structure, an *interior joint* is a point on the structure at which two or more member length axes intersect. For example, in Figure 1.3 points 2, 4, 5, 7, 8, and 10 are interior joints whereas points 1, 3, 6, and 9 are support joints (or exterior joints, or boundary joints).

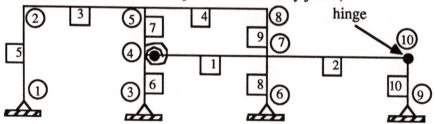

At the joint 4 end of member 1, there is an internal hinge plus
a rotational spring spanning across the hinge.

Figure 1.3 Idealized interior joint conditions

The manner in which the member ends are connected at an interior joint must be accounted for on the line diagram. The types of connections for a structure composed of steel members can be broadly categorized as being one of the following types:

1. *Shear Connection* (or a no moment connection) -- If the connection at joint 10 of Figure 1.3 is as shown in Figure 1.4, it is classified by designers as being a shear connection (or a no moment connection). Thus, an internal hinge is shown on the line diagram at joint 10 of Figure 1.4 to indicate that no moment can be transferred between the ends of members 2 and 10 at joint 10. However, the internal hinge is capable of transferring translational type member end forces (axial forces and shears) between the ends of members 2 and 10 at joint 10. It should be noted that this type of connection can transfer a small amount of moment, but the moment is so small that it can be ignored in design.

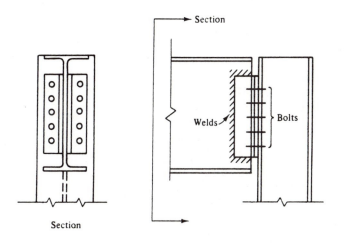

Figure 1.4 Web connection (shear connection)

2. *Rigid Connection* (fully restrained type of construction). If the connection at joint 7 of Figure 1.3 is as shown in Figure 1.5, it is classified by designers as a joint that behaves like a rigid body (or a non-deformable body). Thus, if joint 7 of Figure 1.3 rotates 5 degrees in the counter clockwise direction, the ends of members 1, 2, 8, and 9 at joint 7 also rotate 5 degrees in the counter clockwise direction.

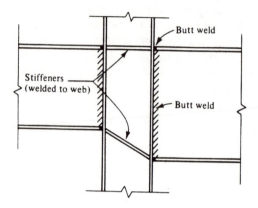

Figure 1.5 Rigid connection--fully welded plus stiffeners

3. *Semi-rigid Connection* (parrtially restrained type of construction). If the beam to column connection at joint 4 of Figure 1.3 is as shown in Figure 1.6, it is classified by designers as being a semi-rigid connection. (Webster's dictionary definition of *semi-rigid* is: rigid to some degree or in some parts.) The web angles fully ensure that the Y- direction displacement at the end of member 1 is identical to the Y-direction displacement of joint 4. (On the line diagram structure in Figure 1.3, joints 3, 4, and 5 lie on the same straight line which is the longitudinal axis of members 6 and 7. Thus, joint 4 is located at the point where the longitudinal axes of members 1, 6, and 7 intersect.) Consequently, joint 4 is treated as being rigid in the Y-direction. However, the top and bottom flange angles in Figure 1.6 are not flexurally stiff enough to ensure that the flanges of member 1 always remain completely in contact with the flanges of members 6 and 7. Thus, joint 4 cannot be treated as being completely rigid. Therefore, at the left end of member 1 in Figure 1.3, a rotational (or spiral) spring is shown to denote that a rotational deformation occurs between joint 4 and the end of member 1. It should be obvious to the reader that a semi-rigid connection is capable of developing more moment than a web connection can develop, but not as much moment as a rigid connection can develop.

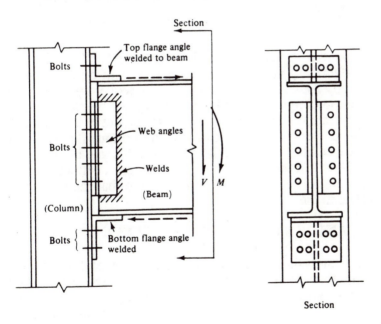

Figure 1.6 Behavior of semirigid connection

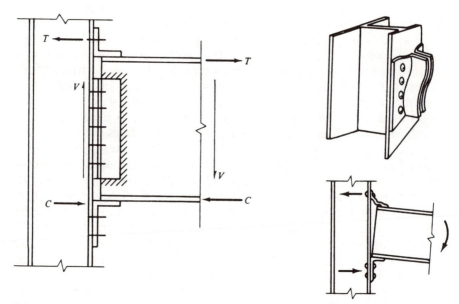

(a) Assumed behavior

(b) Deformation of the connection (separated for clarity)

Figure 1.6 (continued)

In Figure 1.6, the angles are welded to the beam and bolted to the column. M effectively is transferred to the top and bottom flange angles as shown by the dashed lined vectors. Consequently, due to the action of M, the top flange angle and the web angles flexurally deform allowing the top beam flange to translate a finite amount in the direction of the top dashed lined vector. However, the bottom flange angle remains in contact with the column flange. Thus, the gap between the end of the beam end and the column flange is trapezoidal after the angle deformations occur. The bolts resist V and ensure that the beam end does not translate in the Y direction.

1.5 LOADS AND ENVIRONMENTAL EFFECTS

In structural analysis courses the analytical model and the applied loads are given information, and the focus of the analysis courses is on the applicable analysis techniques. In structural design the loads which are to be applied to the analytical model of the structure must be established by the structural designer.

14

Each state in the United States of America has a building code which is mandated by law to be used in the design of an engineered structure. The state building code gives minimum design loads which must be used in the design of a building to ensure a desired level of public safety unless in the structural engineer's judgment higher design loads should be used. Since coping with building codes and determining the applied loads are more appropriately a part of a structural design course dealing with the design of an entire building, the author chooses to give only a brief description of loads and environmental effects in this text. However, he chooses to use the same terminology in the discussion as is used in the building code definitions for the loads and environmental effects.

All loads are treated as being statically applied to the structure and are classified as being one of the following types: *dead loads*; *live loads*; and, *impact loads*. *Environmental effects* due to wind, earthquakes, snow, rain, ice, temperature changes, differential settlement of supports, soil pressures, and hydrostatic pressures are converted into equivalent statically applied live loads.

Examples of *dead loads* are: the weight of the structure; heating, air-conditioning, plumbing, and electrical conduits and fixtures; floor and roof covers; and ceilings. Dead loads do not vary with time in regards to position and weight. Thus, they are not moved once they are in place and, therefore, are called dead loads. A worn floor or roof cover is removed and replaced with a new one in a matter of days. A load which is not there for only a few days in the, say, 50 year life of a structure can be considered a permanent load and classified as a dead load.

Gravity loads which vary with time in regards to magnitude or/and position are called *live loads*. Examples of live loads are: people, furniture, movable equipment, partition walls, file cabinets, and stored goods in general. Forklifts and other types of slow moving vehicles (in a parking garage, for example) may be treated as live loads.

Impact loads are live loads which have dynamic effects that can not be ignored. Examples of impact loads are: cranes, elevators, reciprocating machinery, and vehicular traffic on highway or railroad bridges.

The *effect of an earthquake* on a building is similar to the effect of a football player being clipped (for our purposes, say a clip is a hit around or below the knees and from the blind side). The football player is unaware that he is going to be hit. Consequently, his feet must go in the direction of the person who hits him, but his upper body does not want to move in that direction until the momentum of his lower body tends to drag the upper body in that direction. An earthquake consists of horizontal and vertical ground motions. The horizontal ground motion effect on a structure is similar to the football player being clipped. It is this type of motion that is converted into an equivalent static loading to simulate the effect of an earthquake on a building. An equivalent static loading is applied at all story levels and in the opposite direction of the ground motion since the foundation of the structure remains stationary in a static analysis. It should be noted that *all dynamic loads cannot always be replaced by equivalent static loads*.

The *effects due to wind* are converted into an equivalent static pressure acting on the structure. There is a *basic wind pressure* which is a function of the mass density of air and the wind velocity. This basic wind pressure is given in the building code either as a formula or in tabular form (in pounds per square foot along the height direction of the building). Wind velocity is least at ground level and increases along the height direction of the building. The basic wind pressure is multiplied by a *shape factor* and possibly a *gust factor* to obtain a modified wind pressure which is applied to the structure.

The effects due to *temperature changes* and *differential settlement* of supports are also converted into equivalent static loadings.

1.6 LOAD AND RESISTANCE FACTOR DESIGN

Each state in the USA has a building code which is prepared by a committee of experienced structural engineers. The building code is mandated by law to be used in the design of a public building. The building code defines minimum loads(live, snow, wind) for which the structure must be designed, but the structural designer may use larger loads if they are deemed to be more appropriate. These *service condition loads* are called *nominal loads*. In the Load and Resistance Factor Design(**LRFD**) approach, the nominal loads are multiplied by a *load factor* which is greater than unity in most cases. The factored loads are applied to the structure before performing structural analyses needed in the design process using the LRFD approach. The AISC LRFD Specifications[2] permit either an elastic analysis or a plastic analysis due to the factored loads. Some examples of the load factor are the numerical values shown in the following combined loading cases which the AISC LRFD Specifications require to be investigated to find the critical combination of factored loads:

1) $1.4 \, \mathbf{D}$
2) $1.2 \, \mathbf{D} + 1.6 \, \mathbf{L} + 0.5 \, (\mathbf{L_r} \text{ or } \mathbf{S} \text{ or } \mathbf{R})$
3) $1.2 \, \mathbf{D} + 1.3 \, \mathbf{W} + 0.5 \, (\mathbf{L_r} \text{ or } \mathbf{S} \text{ or } \mathbf{R})$

where: $\mathbf{D}, \mathbf{L}, \mathbf{W}, \mathbf{L_r}, \mathbf{S},$ and $\mathbf{R}$ are *nominal loads*

D is *dead load* due to the weight of the structure and
        permanent features on the structure
L is *live load* due to occupancy and moveable equipment
W is *wind load*
L_r is *roof live load*
S is *snow load*
R is load due to initial *rainwater* or *ice* exclusive
        of the ponding contribution

Cross sectional properties (areas, moments of inertia) given in the AISC Manual are nominal values. The steel mills have tolerances (+ and -) for thicknesses and widths of flanges and webs of a rolled structural steel member. Thus, there is some uncertainty in the cross sectional properties listed in the steel handbook. For a rolled section that is used as a tension member welded to its connections, for example, the limit of internal resistance (*nominal strength*) is the cross sectional area times the yield strength of the steel. If bolted connections are used, fracture of the member in the connection region may be the governing definition of the limit of internal resistance. To account for the uncertainty in the cross sectional area, the resistance (*nominal strength*) is multiplied by a *resistance factor* named ϕ ($\phi < 1$) to obtain the *design strength* of a tension member. In the AISC LRFD Specifications, ϕ (phi) is the symbol for the *resistance factor* (strength reduction factor) and P_n is the *nominal strength* (resistance) for a tension member. If we let P_u be the *required tensile strength* (maximum axial tension force obtained from an elastic factored load analysis), the LRFD Specification requires: $\phi * P_n \geq P_u$. That is, the **resistance factor** times the **nominal tensile strength** **must be equal to or greater than** the **required tensile strength** obtained from an elastic factored load analysis. Some examples of the *resistance factor*, ϕ, are:

1) $\phi = 0.85$ for *axial compression*

2) $\phi = 0.90$ for *shear*

3) $\phi = 0.90$ for *flexure* (bending moment)

4) $\phi = 0.90$ for *tension yielding*

5) $\phi = 0.75$ for *tension fracture*

The load and resistance factors in the AISC LRFD Specifications were developed using a probabilistic approach to ensure that the maximum strength of a structure and each element (member, connection) in the structure is greater than the loads imposed on them, with a reasonable margin of safety. Some of the margin of safety is in the load factors and the other part is in the resistance factors.

In addition to being adequately designed for strength requirements, the structure must perform satisfactorily under nominal or service load conditions. Deflections of floor and roof beams must not be excessive. Relative deflections of the column ends or story drift due to wind load must be controlled. Excessive vibrations cannot be tolerated. Thus, the structural designer must provide a structure that satisfies the owner's performance requirements and the safety requirement on strength as imposed by the applicable building code and LRFD Specifications.

1.7   THE CONSTRUCTION PROCESS

If the framework of the structure is made of steel, the construction process involves the *fabrication, field erection* and *inspection* of the erected structural steel. The general contractor chooses the shop to fabricate the steel and the subcontractor to do the field erection of the steel (in some cases the general contractor erects the steel framework). Field inspection is done by an employee hired by the structural engineer and/or the architect. *Field inspection* is an integral part of the construction process and the final phase of the design process.

Fabrication involves interpreting design drawings and specifications, preparing shop fabrication and field erection drawings, obtaining the material from a steel mill if the needed material is not in the stockpile, cutting, forming, assembling the material into shippable units, and shipping the fabricated units to the construction site.

The fabricator cuts the main members to the correct length, cuts the connection pieces from larger pieces including steel plates, and either punches or drills the holes wherever bolted field connections are specified. A shearing machine is used to cut thin material, and a gas flame torch is used to cut thick material and main members unless extreme precision or a smooth surface is required in which case the cut is made with a saw. If the design specifications do not tolerate as much crookedness in a member as the allowed steel mill tolerances, the fabricator reduces the amount of crookedness by using presses or sometimes by applying heat to localized regions of the member.

The hole punching process may cause minute cracks or may make the material brittle in a very narrow rim around the punched hole. Steel design specifications usually reuire the structural designer to assume that the hole diameter is 1/16 inch larger than the punched hole in order to account for the material which was "damaged" by the punching process.

The steel *field erection* contractor uses his ingenuity and experience to devise an erection plan which involves lifting the fabricated units into place with a crane. Without a proper plan, lifting operations may cause compression forces to occur in members of a truss that were designed to resist only tension, for example. Also, improperly lifting a plate girder could cause local buckling to occur. Temporary bracing generally must be provided by the erection contractor to avoid construction failures due to the lack of three dimensional or space frame stability. After permanent bracing designed by the structural designer, the roof, and the walls are in place, the structure has considerably more resistance to wind loads. Consequently, more failures due to wind loads occur during construction due to the lack of an adequately designed temporary bracing scheme by the erection contractor.

1.8 THE ROLE OF THE STRUCTURAL DESIGNER

Architectural, heating, air conditioning, and other requirements by the owner impose constraints on the structural designer's choice of the structural system for a building. The owner wants a durable, serviceable, and low maintenance structure, and possibly an easily remodeled structure. The structural designer's choice of the structural framing scheme and the structural material for it are influenced by these factors. Sometimes a special architectural effect dictates the choice of the material and framing scheme.

In addition to the obvious possibility of employment as a structural designer for a structural design firm or architectural firm, the following types of employment are possible:
1) *Steel Fabricator* -- detail design of the connections
2) *Government Agencies* -- supervision, inspection, and possibly some design (highway bridges, culverts, piers, overhead sign support structures)
3) *Private Companies* -- some large corporations have their own design and construction teams
4) *Steel Erection* -- prepare erection schemes
5) *Construction Firms* -- design temporary structures

6) *Aerospace Firms* -- detail structural design of components of airplanes, space vehicles, and outer space structures

1.9   SIGNIFICANT DIGITS AND COMPUTATIONAL PRECISION

Dead loads can be estimated very well after all of the member sizes are finalized. However, estimated live loads and equivalent static loads for the effects due to wind and earthquakes are much more uncertain than the estimated dead loads. Perfection is impossible to achieve in the fabrication and erection procedures of the structure.   Also, certain simplifying assumptions have to be made by the analyst to obtain practical solutions. For example, joint sizes are usually assumed to be infinitesimal whereas they really are finite. Interior joints and boundary joints are assumed to be either rigid or pinned whereas they really are somewhere between rigid and pinned. Thus, the final structure is never identical to the one that the structural engineer designed, but the differences between the final structure and the designed structure are within certain tolerable limits.

A digit in a measurement is a significant digit if the uncertainty in the digit is less than 10 units. Standard steel mill tolerance for areas and weights is 2.5% variation. Consider a piece of steel listed in a steel handbook as weighing 100 pounds per foot and having a cross sectional area of 29.4 square inches. The weight variation tolerance is 0.025*100 = 2.5 and the actual weight lies between 97.5 and 102.5 pounds per foot. Since the third digit in 100 is uncertain by only 5 units, the 100 pounds per foot value is valid to three significant digits. However, the area variation tolerance is 0.025*29.4= 0.735 and the actual area lies between 28.665 and 30.135 square inches. Since there are 14.7 units of uncertainty in the third digit of 29.4, only the first two digits in the value 29.4 are significant and a recorded value of 29.0 would be more appropriate in the steel handbook.

Most computers will accept arithmetic constants having an absolute value in the range 1×10^{-35} to 1×10^{35}. Many computers will accept a much wider range. A computer holds numeric values only to a fixed number of digits, usually the equivalent of between 6 and 16 decimal (or base 10) digits. The number of decimal digits held is called the precision of the arithmetic constant.

For discussion purposes, suppose that the loads and structural properties are accurate to only 3 significant digits. Most commercial structural analysis programs use at least 16 digit precision in the solution of the set of simultaneous equations. In a multistory building, there may be hundreds or thousands of equations in a set. If only 3 digit precision were used in computerized solutions, the truncation and round off errors in the mathematics would in some cases contribute much more uncertainty in the computed results than the structural engineer has in the loads and structural properties.

Reconciliation of the actual number of significant digits in the computed results should be made by the structural designer after all of the computed results are available. The author always makes his electronic calculator computations using the maximum available precision in the same manner that a computer would make them in floating point form. If the loads have units of either kips or kilo-newtons, the author records the computed reactions and final member end forces rounded in the first digit to the right of the decimal in fixed decimal form.

Chapter 2

STRUCTURAL STEEL

2.1 COMPOSITION AND TYPES

Steel is used to produce a variety of products ranging from paper clips to space vehicles and is extensively used for frameworks of buildings, bridges, buses, cars, conveyors, cranes, platforms, pipelines, storage tanks, towers, trucks and other structures.

Prior to about 1960 steel used in building frameworks was ASTM(American Society for Testing and Materials) designation **A7** with a yield strength of 33 ksi. Today there are a variety of ASTM designations available with yield strengths ranging from 24 ksi to 100 ksi. *Yield strength* is the term used to denote the *yield point* (see Figure 2.1) of the common structural steels or the stress at a certain offset strain for steels not having a well defined yield point.

Steel is composed almost entirely of iron, but contains small amounts of other chemical elements to produce desired physical properties such as *strength, hardness, ductility, toughness*, and *corrosion resistance*. Carbon is the most important of the other elements. Increasing the carbon content produces an increase in strength and hardness, but decreases the ductility and toughness. Manganese, silicon, copper, chromium, columbium, molybdenum, nickel, phosphorus, vanadium, zirconium, and aluminum are some of the other elements that may be added to structural steel. Hot rolled structural steels may be classified as *carbon steels, high strength low alloy steels*, and *alloy steels*.

Carbon steels contain the following maximum percentages of elements other than iron: 1.7% carbon, 1.65% manganese, 0.60% silicon, and 0.60% copper. Carbon and manganese are added to increase the strength of the pure iron. Carbon steels are divided into four categories: 1) *low carbon*(less than 0.15%); 2) *mild carbon* (0.15 to 0.29%); 3) *medium carbon* (0.30 to 0.59%); and, 4) *high carbon* (0.60 to 1.70%). Structural carbon steels are of the mild carbon category. For example, **A36** with a yield strength of 36 ksi is the most common structural steel and has a maximum carbon content of 0.25 to 0.29% depending on the thickness. Structural carbon steels have a distinct yield point as shown in Figure 2.1 on curve (a). The carbon steels of Table 2.1 are A36, A53, A500, A501, A529, A570, and A709(Grade 36) with yield strengths ranging from 25 to 100 ksi.

High strength low alloy steels (see curve (b) of Figure 2.1) have a distinct yield point ranging from 40 to 70 ksi. Alloy elements such as chromium, columbium, copper, manganese, molybdenum, nickel, phosphorous, vanadium, and zirconium are added to improve some of the mechanical properties of steel by producing a fine instead of a coarse microstructure obtained during cooling of the steel. The high strength low alloy steels of Table 2.1 are A242, A441, A572, A588, A606, A607, A618, and A709(Grades 50 and 50W).

Alloy steels (see curve (c) of Figure 2.1) do not have a distinct yield point. Their yield strength is defined as the stress at an offset strain of 0.002 with yield strengths ranging from 80 to 110 ksi. These steels generally have a maximum carbon content of about 0.20% to limit the hardness that may occur during heat treating and welding. Heat treating consists of quenching (rapid cooling with water or oil from 1650 °F to about 300 °F) and tempering (reheating to 1150 °F and cooling to room temperature). Tempering somewhat reduces the strength and hardness of the quenched material but significantly improves the ductility and toughness. The quenched and tempered alloy steels of Table 2.1 are A514 and A709(Grades 100 and 100W).

Threaded fasteners and *bolts* are composed of steel designated as:

1) **A307**(low carbon) -- A307 bolts, commonly referred to as machinebolts, do not have a distinct yield point(minimum yield strength of 60 ksi is taken at a strain of 0.002), and are not allowed by AISC Specifications in buildings except for temporary construction and where strength is not important.

2) **A325**(medium carbon; quenched and tempered with not more than 0.30% carbon) -- Bolts have a 0.2% offset minimum yield strength of 92 ksi (0.5 to 1 inch diameter bolts) and 81 ksi (1.125 to 1.5 inch diameter bolts) and an ultimate strength of 105 to 120 ksi.

3) **A449 bolts** have tensile strengths and yield strengths similar to A325 bolts, have longer thread lengths, and are available up to 3 inches in diameter. A449 bolts and threaded rods are permitted only where greater than 1.5 inch diameters are needed.

4) **A490 bolts** are quenched and tempered, have alloy elements in amounts similar to A514 steels, have up to 0.53% carbon, and a 0.2% offset minimum yield strength of 115 ksi(2.5 to 4 inch diameter) and 130 ksi(less than 2.5 inch diameter).

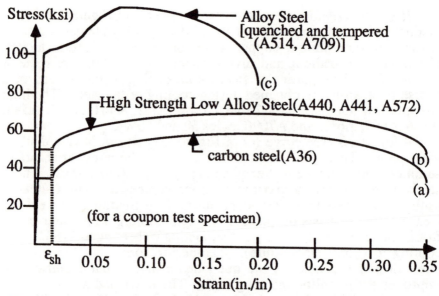

Figure 2.1 Typical Stress-Strain curves for the 3 classes of steels

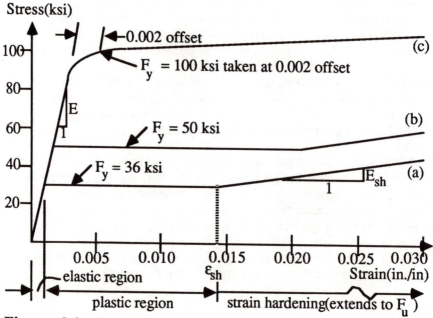

Figure 2.2 Enlargement of Figure 2.1 in vicinity of yield point

Table 2.1 Steels Used for Buildings and Bridges

ASTM Designation	F_y(ksi)	F_u(ksi)	Maximum thickness (in.)	Common Usage
A36	32	58-80	over 8	General; Buildings
	36	58-80	to 8	ditto
A53	30	48		Pipe
	35	60		ditto
A242	42	63	1.5 to 4	Bridges
	46	67	0.75-1.5	ditto
	50	70	to 0.75	ditto
A440 (same as A242)				Bolted construction
A441	40	60	4 to 8	Welded construction
	42	63	1.5-4	ditto
	46	67	0.75-1.5	ditto
	50	70	to 0.75	ditto
A500	33,42,46	45,58,62	Round	Cold-formed tubing
	39,46,50	45,58,62	Shaped	ditto
A501	36	58		Tubing
A514	90	100-130	2.5 to 6	Plates for welding
	100	110-130	to 2.5	ditto
A529	42	60-85	to 0.5	Pre-engineered frames
A570	25,30,33	45,59,52		Cold-formed sections
	40,42	55,58		ditto
A572	42	60	to 6	Buildings; Bridges
	50	65	to 2	Buildings; Bridges
	60,65	75,80	to 1.25	Buildings

Table 2.1 Continued

ASTM Designation	F_y(ksi)	F_u(ksi)	Maximum thickness (in.)	Common Usage
A588	42	63	5 to 8	Weathering Steel
	46	67	4 to 5	ditto
	50	70	to 4	ditto(for bridges)
A606	45,50	65,70		Cold-forming Sheets
A607	45,50,55	60,65,70		Cold-forming Sheets
	60,65,70	75,80,85		ditto
A611	25,30,33	42,45,48		Cold-forming Sheets
	40,80	52,82		ditto
A618	50	65,70		Hot-formed tubing
A709	see:A36,A441(50ksi),A588(50W),A514(100)			Bridges

Weld electrodes are classified as E60XX, E70XX, E80XX, E90XX, E100XX, and E110XX where: E denotes electrode; the first two digits denote the tensile strength in ksi; and XX represent numbers indicating the usage of the electrode.

2.2 MANUFACTURING PROCESS

At the steel mill, the manufacturing process begins at the blast furnace where iron ore, limestone and coke are dumped in at the top and molten pig iron comes out at the bottom. Then the pig iron is converted into steel in basic oxygen furnaces [electric furnaces (newer mills) and open hearth furnaces(older mills)]. Oxygen is essential to oxidize the excess of carbon and other elements and must be highly controlled to avoid gas pockets in the steel ingots since gas pockets will become defects in the final rolled steel product. Silicon and aluminum are deoxidizers used to control the dissolved oxygen content. Steels are classified by the degree of deoxidation: 1) *killed steel* (highest dexoxidation); 2) *semi-killed steel* (intermediate deoxidation); and *rimmed steel* (lowest deoxidation).

Potential mechanical properties of steel are dictated by the chemical content, the rolling process, finishing temperature, cooling rate, and any subsequent heat treatment. In the rolling process, material is squeezed between two rollers revolving at the same speed in opposite directions. Thus, rolling produces the steel shape, reduces it in cross section, elongates it, and increases its strength. Ordinarily, ingots are poured from the basic oxygen furnaces, reheated in a soaking pit, rolled into slabs, billets, or blooms in the bloom mill, and then rolled into shapes, bars, and plates in the breakdown mill and finishing mill. If the continuous casting process is used, the ingot stage is bypassed.

A chemical analysis, also known as the heat or ladle analysis, is made on samples of the molten metal and is reported on the mill test certificate for the heat or lot(50 to 300 tons) of steel taken from each steel making unit. One to eight hours are required to produce a heat of steel depending on the type of furnace being used.

Mechanical properties(*modulus of elasticity, yield strength, tensile strength*, and *elongation* to determine the *degree of ductility*) of steel are determined from tensile tests of specimens taken from the final rolled product. These mechanical properties listed on the mill test certificate normally exceed the specified properties by a significant amount and merely certify that the test certificate meets prescribed steel making specifications. Each piece of steel made from the heat of steel covered by the mill test certificate does not have precisely the properties listed on the mill test certificate. Therefore, structural designers do not use the mill test certificate properties for design purposes. The minimum specified properties listed in the design specifications are used by the structural designer.

2.3 STRENGTH AND DUCTILITY

Strength and *ductility* are important characteristics of structural steel in the structural design process. Suppose identical members(same length and same cross sectional area) are made of wood, reinforced concrete, and steel. The steel member has the greatest strength which permits designers to use fewer columns in long clear spans of relatively small members to produce steel structures with minimum dead weight.

Ductility, the ability of a material to undergo large deformations without fracture, permits a steel member to yield when overloaded and redistribute some of its loads to other adjoining members in the structure. Without adequate ductility: 1) there is a greater possibility of a fatigue failure due to repeated loading and unloading of a member; and, 2) a brittle fracture can occur.

Strength and *ductility* are determined from data taken during a standard, tensile, load-elongation test. [The author contends that more appropriately for a member subjected to bending, the area under the Moment-Curvature curve is a better measure of ductility due to bending.] A stress-strain curve such as Figure 2.1 can be drawn using the load-elongation test data. On the stress-strain curve, after the peak or *ultimate strength*, F_u, is reached, a descending branch of the curve occurs for two reasons:

1) *mathematical* -- stress is computed as the applied load divided by the original, unloaded, cross sectional area whereas the actual cross sectional area reduces rapidly after the ultimate strength is reached; and, 2) *due to the load being hydraulically applied in the lab* -- if the load were applied by pouring beads of lead into a bucket, for example, no decrease in load would occur from the time the ultimate strength was obtained until fracture occurred and a horizontal, straight line would occur on the usual stress-strain curve from the ultimate strength point to the fracture point.

2.4 CROSS SECTIONAL PROPERTIES

Figure 2.3 shows a cross section of a steel rolled shape designated as a W section which is the most common shape used in structural steel design as a *beam* (bending member), a *column* (axial compression member), and a *beam-column* (axial compression plus bending member).

Suppose Figure 2.3 is the cross section of a column of length L. Let A denote the cross sectional area of Figure 2.3 and P denote the axial compression force applied to each end of the column(see Figure 2.4). In Figure 2.3, the cross section is composed of five *compression elements* each of which is subjected to an uniform compressive stress of P/A. Compression elements of a cross section are classified as being either *stiffened* or *unstiffened* (**projecting**). A *stiffened compression element* is attached on both ends to other compression elements. An *unstiffened compression element* is not attached to anything on one end and is attached to another compression element(s) on the other end.

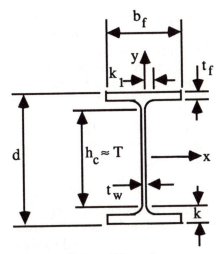

Figure 2.3 W section

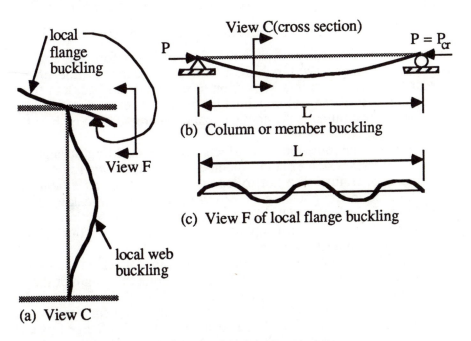

(c) View F of local flange buckling

(a) View C

Figure 2.4 Local buckling and member buckling

Essentially, a W section shape(Figure 2.3) is in the form of the capital letter **I** which is composed of two horizontal lines and one

29

vertical line. The vertical line is called the *web* and the horizontal lines are called *flanges*. Thus, the web is a stiffened compression element since it is attached on both ends to other elements of the cross section. Also, each flange(there is a top flange and a bottom flange) is composed of two unstiffened compression elements since each flange is attached to one end of the web and the other ends of each flange are not attached to anything.

Each compression element of the cross section in Figure 2.3 essentially is a rectangle. The longer side of the rectangle is the *width* and the other side is the *thickness*. Each compression element has a property known as the **width-thickness ratio** or $\frac{b}{t}$ in mathematical terms.   For each of the four unstiffened elements in Figure 2.3, $b = 0.5b_f$ and $t = t_f$ where b_f is the overall or total width of each top and bottom flange and t_f is the flange thickness. For the stiffened element (the web) in Figure 2.3, $b = h_c$ and $t = t_w$ where t_w is the thickness of the web and $h_c = d - 2*(t_f + k_1 - 0.5t_w)$ is the clear height of the web.

If $\frac{b}{t}$ of a compression element is too large, the compression element buckles locally as shown in Figure 2.4 before the column buckles. Consequently, the AISC LRFD Specifications limit b/t of compression elements in a W section to prevent local buckling from occurring before the column buckles in which case the buckling strength of the column is the nominal axial compressive strength of the member.

Another phenomenon that affects column buckling strength is *residual stresses* which exist in a member due to: 1) the *uneven cooling* to room temperature of a hot rolled steel product; 2) *cold bending* (process used in straightening a crooked member and in making cold-formed steel sections); and, 3) *welding* two or more sections/plates together to form a built up section(for example, four plates interconnected to form a box section).

Consider a hot rolled W section product after it leaves the rollers for the last time. Consider any cross section along the length of the W section product. The flange tips and the middle of the web cool under room temperature conditions at a faster rate than the junction regions of the flanges and the web. Steel shrinks as it cools. The flange tips and the middle of the web shrink freely when they cool since the other regions of the cross section have yet to cool. When the junction regions of the flanges and the web shrink, they are not completely free to shrink since they are interconnected to the flange tips and the middle of the web regions which have already cooled. Thus, the

30

last to cool regions of the cross section contain residual tensile stresses whereas the first to cool regions of the cross section contain residual compressive stresses. These residual stresses, caused by shrinkage of the last to cool portions of the cross section and their being interconnected to regions which are already cool, have a symmetrical pattern with respect to the principal axes of the cross section of the W section. Therefore, the residual stresses are self equilibrating and do not cause any bending about either principal axis of a cross section at any point along the length direction of the member. Residual stresses in a W section are in the range of 10 to 15 ksi regardless of the yield strength of the steel.

If a W section column buckles inelastically, the first to cool regions of the cross section yield when the residual compressive stress due to cooling plus the applied compressive stress due to the buckling load reaches the yield strength of the steel. However, the last to cool regions of the cross section contain residual tensile stresses plus the applied compressive stress due to the buckling load. Consequently, some portions of these last to cool regions of the cross section are still elastic when inelastic column buckling occurs.

2.5 OTHER PROPERTIES

For purposes of most structural design calculations, the following values are used where applicable for steel:
1) Weight = 490 pounds per cubic foot
2) Coefficient of Thermal Expansion, **CTE** = 0.0000065 strain/°F)
3) Poisson's ratio, $\upsilon = 0.3$

The stress-strain curves shown in Figure 2.1 are for room temperature conditions. As shown in Figure 2.5, after steel reaches a temperature of about 200 °F, the yield strength, tensile strength, and modulus of elasticity are significantly influenced by the temperature of the steel. Also, at high temperatures steel creeps (deformations increase with respect to time under a constant load). Temperatures in the range shown in Figure 2.5 can occur in members of a building in case of a fire, in members over an open flame in a foundry, and in the vicinity of field welds, for example.

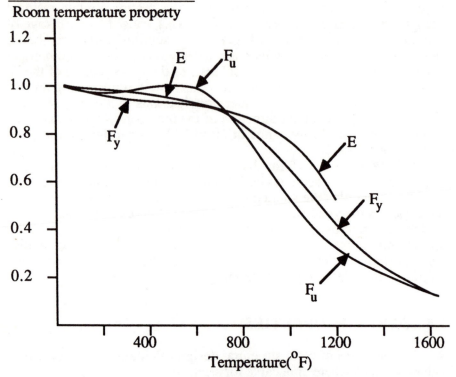

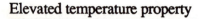

NOTE: For temperatures below 32°F, the properties shown increase. However, ductility and toughness decrease.

Figure 2.5   Effect of temperature on properties of steel

Temperature and prior straining into the strain hardening region have an adverse effect on ductility. Fractures at temperatures significantly below room temperature are brittle instead of ductile. *Toughness* (ability to absorb a large amount of energy prior to fracture) is related to ductility. Toughness usually is measured in the lab by a Charpy V-Notch impact test in which a standard notched specimen chilled or heated to a specified temperature is struck by a swinging pendulum. Toughness, as implied by the type of test for toughness, is important for structures subjected to impact loads(earthquakes, vertical motion of trucks on bridges, and sudden stops on elevator cables). Killed steels and heat treated steels have the most toughness. As shown in Figure 2.6, ductility is significantly reduced after a structure

32

has been overloaded into the strain hardening region. Overloaded was chosen by the author as the descriptor since a building framework does not experience strains in the strain hardening region under normal service condition loads except for severe earthquakes, for example. However, corners (bends of 90 degrees or more at room temperature) of cold-formed steel sections are strained into the strain hardening range.

Corrosion(rusting) resistance increases as the temperature increases up to about 1000 °F. In the welding process a temperature of about 6500 °F occurs at the electric arc tip of a welding electrode. Thus, high temperatures due to welding occur and subsequently dissipate in a member in the vicinity of welds. High strength low alloy steels have several times more resistance to rusting than carbon steels. *Weathering steels* form a thick crust of rust which protects the structure from further exposure to oxidation.

Weldability (relative ease of producing a satisfactory, crack free, structurally sound joint) is an important factor in structural steel design since most connections in the fabrication shop are made by welding using automated, high speed welding procedures wherever possible. The temperature of the electric arc increases as the speed of welding increases and more of the structural steel mixes with the weld. Thus, some steels are more suited to high speed welding than others.

Members and their connections in a highway or railway bridge truss, for example, may be repeatedly loaded and unloaded millions of times during the life of the bridge. Some of the diagonal truss members may be in tension and later on in compression as a truck traverses the bridge. Even if the yield point of the steel in a member or its connections is never exceeded during the repeated loading and unloading occurrences, a fracture can occur and is called a *fatigue fracture*. Anything that reduces the ductility of the steel in a member or its connections increases the chances of a brittle, fatigue fracture. Thus, *fatigue strength* may dictate the definition of nominal strength of members and connections that are repeatedly loaded and unloaded a very large number of times during the life of the structure. Indeed, the life of a repeatedly loaded and unloaded structure may be primarily dependent on the fatigue strength of its members and connections.

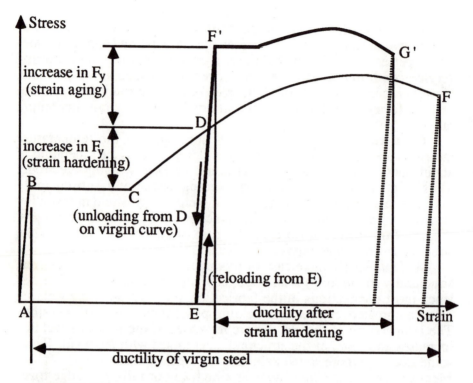

A,B,C,D,F is the virgin steel curve
D,E,D,F is unloading and immediate reloading curve
D,E,F',G' is unloading and reloading considerably later curve

Figure 2.6 Stress-Strain curves

Chapter 3

TENSION MEMBERS

3.1   INTRODUCTION

A tension member is designed on the assumption that the member has to provide only an axial tensile strength. Cables or guy wires are used as tension members to stabilize wood poles that support telephone and electricity transmission lines. Steel cables(wire ropes) and very slender rods(length/diameter 300) have negligible bending stiffness. Thus, the assumption that a cable or very slender rod only provides tensile strength is indeed very reasonable. A tension member in a truss is fastened by welds or bolts at the member ends to either other members or connection plates(gusset plates). Truss members do not necessarily have negligible bending stiffness. Therefore, if the structural designer wants a structure to behave like a truss(all joints assumed to be pins and no bending occurs in any member), the design details must be chosen such that negligible bending occurs in each member. This means that the design details must: 1) provide for all loads except the self weight of the members to actually occur only at truss joints; and, 2) ensure that the joints do not cause appreciable member end moments to occur. The author's point is -- if a structural designer wants a structure to behave in a certain manner, the structural details must be carefully chosen such that the desired structural behavior is closely approximated.

In the LRFD approach which was briefly described in Section 1.6, the members and connections of a structure are designed to have adequate strength to resist the factored loads imposed on the structure. For a member or a connection, the design strength is the nominal strength times a resistance factor. Thus, for each member and each connection, the design strength must be greater than or equal to the required strength determined from a factored load analysis.

After the structure has been adequately *designed for strength*, the structural designer investigates the performance of the structure under service conditions. If a tension member is too flexible (does not have enough bending stiffness): 1) special handling in the fabrication shop and in the field erection stages may be necessary resulting in extra costs; 2) the member may sag excessively due to its own weight; and, 3) in a bridge truss exposed to wind or inside a building containing large machines with rotating parts, the member may vibrate too much.

Thus, in addition to adequate *strength*, a member and the entire structure must have adequate *stiffness* for serviceability reasons. Many of the owner's serviceability requirements can be met by ensuring that deflections do not exceed acceptable limits. Some of the common serviceability problems are[3]:

1. Local damage of nonstructural elements (for example, windows, ceilings, partitions, walls) occurs due to displacements caused by loads, temperature changes, moisture, shrinkage, and creep.
2. Equipment (for example, elevators) does not function normally due to excessive displacements.
3. Drift or/and gravity direction deflections are so noticeable that occupants become alarmed.
4. Extensive nonstructural damage occurs due to a tornado or a hurricane.
5. Structural deterioration occurs due to age and usage (for example, deterioration of bridges and parking decks due to de- icing salt).
6. Motion sickness of the occupants occurs due to excessive floor vibrations caused by routine occupant activities or lateral vibrations due to the effects of wind or an earthquake.

These serviceability problems can be categorized as a function of either the gravity direction deflection or the lateral deflection.

Let: L = span length of a floor or roof member; h = story height.

Deflection Index	Typical Serviceability Behavior
$h/1000$	Not visible cracking of brickwork
$h/500$	Not visible cracking of partition walls
$h/300$ or $L/300$	Visible architectural damage Visible cracks in reinforced walls Visible ceiling and floor damage Leaks in the structural facade
$L/200$ to $L/300$ $h/200$ to $h/300$	Cracks are visually annoying Visible damage to partitions and large, plate glass windows
$L/100$ to $L/200$ $h/100$ to $h/200$	Visible damage to structural finishes Doors, windows, sliding partitions, and elevators do not function properly

It is customary steel design practice to limit the Deflection Index to:
1. L/360 *due to live load* on a floor or snow load on a roof when the beam supports a plastered ceiling.
2. L/240 *due to live load* or *snow load* if the ceiling is not plastered.
3. h/667 to h/200 for each story *due to the effects of wind* or *earthquakes* -- only a range of limiting values can be given for many reasons (type of facade, activity of the occupants, routine design, innovative design, structural designer's judgment and experience).
4. H/715 to H/250 for entire building height H *due to the effects of wind* or *earthquakes* -- comment in item 3 applies here too.

It is important to note that: the Deflection Index limits for drift are about the same as the accuracy which can be achieved in the erection of the structure; and, the largest tolerable deflection due to live load is 0.5% of the member length. Consequently, deflections are grossly exaggerated for clarity on deflected structure sketches in textbooks.

The first undergraduate steel design course deals almost exclusively with individual members in regards to behavior, design strength definitions, and member sizing(selecting a member with adequate design strength). A subsequent structural design course deals with all aspects of design for an entire building. Since the material in this text was prepared for the first steel design course, the remainder of this chapter deals with design considerations for individual tension members. However, the nominal strength definition of a tension member is different for welded and bolted connections at the member ends. Consequently, some discussion of the member end connections is necessary.

To facilitate the discussion of connections and to illustrate how a member's required strength is determined from a factored load analysis, the author chooses to use the plane frame structure shown in Figures 3.1 and 3.2[see text Appendix A for the factored load analysis results]. This structure is a roof truss supported by two beam-columns (members 1 to 4 in Figure 3.2). A beam-column is a member that is subjected to axial compression plus bending. The behavior and design of beam-columns are discussed in Chapter 6. In Figure 3.2, members 1 to 4 and the roof truss ends are interconnected to provide resistance due to wind as well as overall lateral stability of the structure for the gravity direction loads. To provide resistance due to wind perpendicular to the plane of Figure 3.2, some bracing scheme(see Figure 3.3 for an example) must be devised and designed.

Since subsequent discussions(in this and other chapters) of the structure in Figure 3.2 will be related to connections and required strength, the following examination of displacements for serviceability purposes is presented now. From text Appendix A, due to nominal (service condition) loads:

1) at joint 2, *due to wind only*, the deflection is:
   1.64 inches = 0.137 feet = (h/154) where h = 2l feet

   According to item 3 on page 3-2, a deflection index of h/154 will not be acceptable since h/154 does not lie in the range of h/667 to h/200. Members 1 to 4 in the analysis for Figure 3.2 are too flexible and their moment of inertia needs to be multiplied by about two or more(this conclusion was made after looking at the following information).

2) at joint 12 *due to live loads* (snow plus the crane loads) only, the vertical deflection is:
   1.18 inches = 0.099 feet = (L/608) where L = 60 feet

   According to item 2 on page 3-2, the *live load deflection* should not exceed (L/240) = 0.25 feet and 0.099 ft. is lessthan 0.25 ft. Consequently, based on the member properties used for the analysis of Figure 3.2, the truss has more than adequate stiffness for gravity loads.

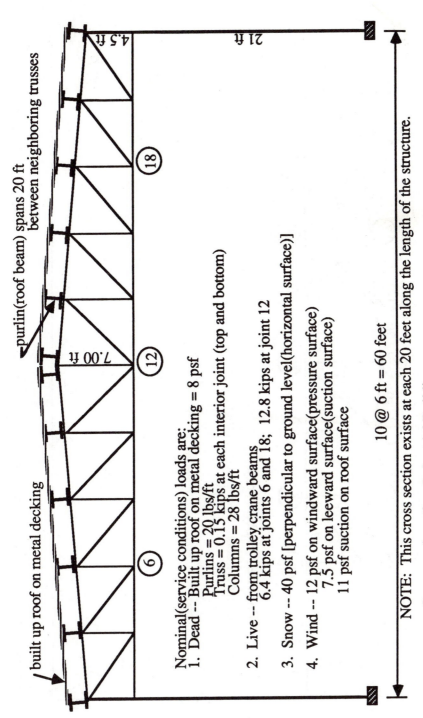

built up roof on metal decking

purlin(roof beam) spans 20 ft
between neighboring trusses

4.5 ft

21 ft

7.00 ft

18

12

6

Nominal(service conditions) loads are:
1. Dead -- Built up roof on metal decking = 8 psf
          Purlins = 20 lbs/ft
          Truss = 0.15 kips at each interior joint (top and bottom)
          Columns = 28 lbs/ft

2. Live -- from trolley crane beams
          6.4 kips at joints 6 and 18;   12.8 kips at joint 12

3. Snow -- 40 psf [perpendicular to ground level(horizontal surface)]

4. Wind -- 12 psf on windward surface(pressure surface)
           7.5 psf on leeward surface(suction surface)
           11 psf suction on roof surface

10 @ 6 ft = 60 feet

NOTE: This cross section exists at each 20 feet along the length of the structure.

Figure 3.1 Cross section of an Industrial Building

39

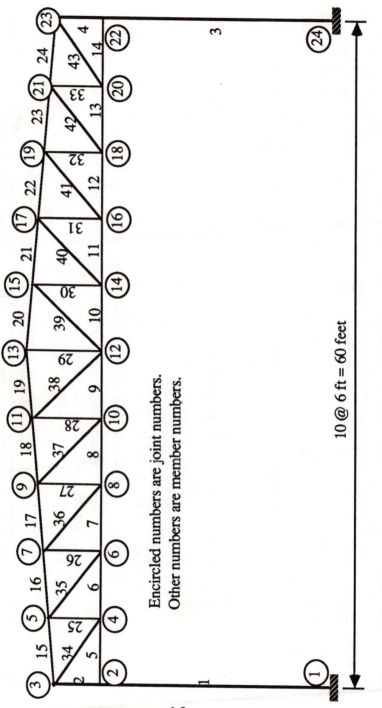

Encircled numbers are joint numbers.
Other numbers are member numbers.

10 @ 6 ft = 60 feet

Figure 3.2 Joint numbers, member numbers, and load cases for structure shown in Figure 3.1

40

Nominal(service conditions) Loadings are:

(1) Dead -- 1.51 kips down at joints 5,7,9,11,15,17,19,21
        1.92 kips down at joint 13
        0.80 kips down at joints 3 and 23
        0.70 kips down at joints 2 and 22
        0.15 kips down at joints 4,6,8,10,12,14,16,18,20

(2) Live -- 6.40 kips down at joints 6 and 18 | crane loads
        12.8 kips down at joint 12

(3) Snow -- 4.80 kips down at joints 5,7,9,11,15,17,19,21
        2.40 kips down at joints 3 and 23

(4) Wind -- 0.24 k/ft to right on members 1 and 2
        0.15 k/ft to right on members 3 and 4
        0.11 kips to left and 1.32 kips up at joints 5,7,9,11
        0.11 kips to right and 1.32 kips up at joints 15,17,19,21
        1.32 kips up at joint 13
        0.055 kips to left and 0.66 kips up at joint 3
        0.055 kips to right and 0.66 kips up at joint 23

    due to suction on the roof

LRFD Loadings which must be considered are: (see LRFD page 6-25)

(5) 1.4D

(6) 1.2D + 1.6L + 0.5(L or S or R)

(7) 1.2D + 1.6(L or S or R)

(8) 1.2D + 1.3W + 0.5L + 0.5(L or S or R)

(9) 0.9D + 1.3 W

Figure 3.2   (continued)

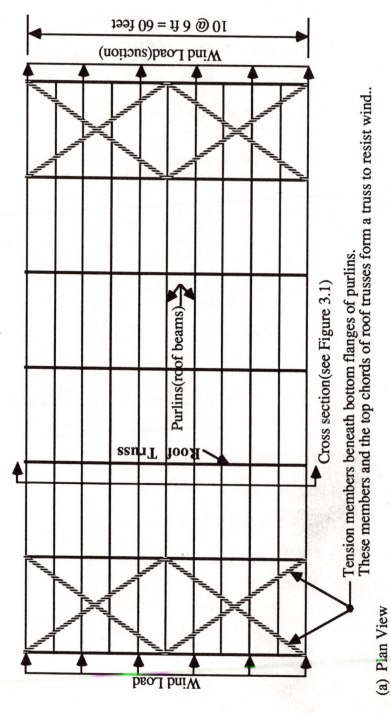

10 @ 6 ft = 60 feet

Wind Load(suction)

Cross section(see Figure 3.1)

Purlins(roof beams)

Roof Truss

Tension members beneath bottom flanges of purlins.
These members and the top chords of roof trusses form a truss to resist wind.

Wind Load

(a) Plan View

Figure 3.3   Side elevation view and plan view of building

42

3.2 EFFECT OF RESIDUAL STRESSES

Consider a laboratory tension test of a particular W section member. As described in Section 2.4, there are residual stresses in the member due to the uneven cooling process of a hot rolled member. As shown in Figure 3.4a: the *maximum residual compressive stress*, f_{rc}, occurs at the flange tips and at midheight of the web; and, the *maximum residual tensile stress*, f_{rt}, occurs at the junction of the flanges and web. The residual stresses vary through the thickness of the flanges and web. It should be noted that cross sectional geometry (flange thickness and width; web thickness and depth) influences the cooling rate and residual stress pattern. Some W sections are configured to be efficient as columns(axial compression members) and other W sections are configured to be efficient as beams(bending members). Depending on the cross sectional geometry, some W sections have only residual tensile stresses in the web with the maximum value occurring at the junction of the flanges and web. Furthermore, the magnitude of the residual stresses is smaller for quenched and tempered members. Thus, the residual stress pattern as well as average values of f_{rc} and f_{rt} through the thickness are dependent on several variables. Residual stress magnitudes on the order of 10 to 15 ksi or more occur if the member is not quenched and tempered.

In a tensile test of a W section member, as shown in Figure 3.4c: 1) fibers in the cross section begin to yield when the applied tensile stress(T/A) and the residual tensile stress add up to the yield stress; and, 2) all fibers in the cross section yield before the first to yield fibers begin to strain harden. Thus, the item 2) phenomenom is the same condition that occurs in a coupon test. Consequently, the yield strength of a tension member is not affected by the presence of residual stresses. However, the fatigue strength of a tension member is affected by the presence of residual stresses.

3.3 TENSION MEMBER WITH WELDED CONNECTIONS

In the following discussion, the assumptions are: 1) there are no holes in the member; and, 2) the member ends are fastened by welds to joints. The AISC LRFD Specification definition of *design strength* is *a resistance factor times the nominal strength*. Since the resistance factor is less than unity, a resistance factor is a strength reduction factor. Separate design strength definitions are given for members, welds, and joints. The governing design strength for a tension member

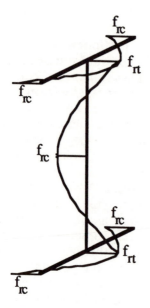

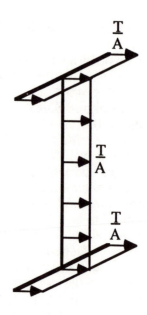

(a) Residual stresses

(b) Tensile test stress

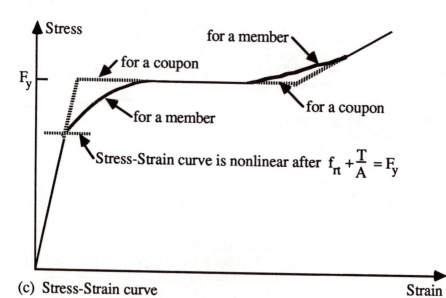

(c) Stress-Strain curve

Figure 3.4 Tensile test of a W section member

welded at the member ends to joints is the least design strength value of
the: 1) member; 2) welds; and, 3) joints. The definitions of design
strength for welds and joints are given later in this chapter.

Consider Figure 3.5 which shows a tension member fastened by
fillet welds to a gusset plate(connection plate). Along the member at
some finite distance from the welds, as shown in Section 3.2 all cross
sectional fibers can attain the yield strength provided the welds and
gusset plate are stronger than the member. In the region of the member
end connection, the stress distribution due to the applied load is not
uniform in the member since the edge of one leg of the angle section in
Figure 3.5c is not welded to the gusset plate. Therefore, a transition
region exists from the connection region to some finite distance from
the connection where the stress distribution in the member becomes
uniform when yielding occurs. Thus, before yielding occurs in the
member, the connection region of the member end usually experiences
strain hardening and fracture can possibly occur in the region where
strain hardening occurs.

If all cross sectional fibers of a member yield in tension, the
member elongates excessively which can precipitate failure somewhere
in the structural system of which the tension member is a part. *Fracture
is a failure condition.* Consequently, for a tension member that does
not contain any holes, the design strength is the least ϕP_n value
obtained from:
1) yielding in the gross section (see Figure 3.5b)
   [LRFD D1 page 6-36]

   $\phi P_n = 0.90*F_y*A_g$
   $A_g = 3.5*0.25 + (3 - 0.25)*0.25 = 1.56$ in^2 for Figure 3.5b
   $\phi P_n = 0.90*36*1.56 = 50.5$ kips
2) fracture in the member end (see Figure 3.5c)
   [LRFD D1 page 6-36]
   [LRFD B3 page 6-29]

   $\phi P_n = 0.75*F_u*A_e$
   where, for Figure 3.5c, $A_e = U*A_g$
   U is a reduction coefficient(see LRFD B3 page 6-30).
   For Figure 3.5c, U = 0.85 from item b of LRFD B3.

   $\phi P_n = 0.75*58*0.85*1.56 = 57.7$ kips
For each LRFD Specification cited by the author, there may be useful
information in the LRFD Commentary. For example, LRFD page 6-
147 gives: $U = 1 - \bar{x}/L$ where L is the length of the connection and $\bar{x}$
is the distance from the connection plate to the centroid of the member

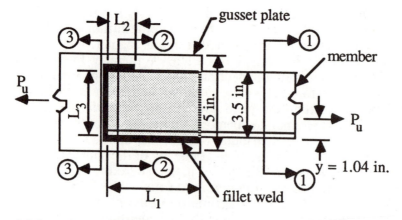

A36 steel[see LRFD page 6-123: F_y = 36 ksi; F_u = 58 ksi]
Gusset plate: 5/16 by 5
Member: L3.5X3X1/4(see LRFD page 1-56,57)
Fillet welds: 3/16 inch E70 electrodes[F_y = 70 ksi]
$$L_1 = 6.75 \text{ in.; } L_2 = 1.85 \text{ in.; } L_3 = 3.50 \text{ in.}$$

(a) Member end connection detail

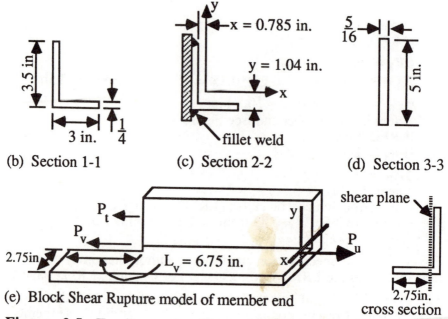

(b) Section 1-1   (c) Section 2-2   (d) Section 3-3

(e) Block Shear Rupture model of member end   2.75in.
cross section

Figure 3.5 Tensile member fillet welded to a gusset plate

46

part being connected. For Figure 3.5c, U = 1 - 0.785/6.75 = 0.884 which could be used instead of the U = 0.85 value which was used.

The author wonders why the last paragraph of LRFD Spec. D1 on page 6-36 contains: "or the effective area of the welds as defined in Sect. J2". There is no explanation in the LRFD Commentary for the inclusion of the phrase enclosed in quotation marks. The author believes he has correctly accounted for this phrase in computing the design strength of the fillet welds (see pages 47,48).

3) block shear rupture of the member end (see Figure 3.5e)
   [LRFD J4 page 6-72; Commentary page 6-187]
   LRFD Commentary page 6-187 should be used for tension members. LRFD Eqn(J4-1)[page 6-73] is used as the first term in LRFD Eqn(C-J4-2)[page 6-187]. There are two possible modes of block shear rupture in Figure 3.5e:
   1. fracture on tensile plane and yielding on shear plane
      $P_1 = 0.75*[(\text{tensile fracture force}) + (\text{shear yield force})]$
      $P_1 = 0.75*[F_u*t*w + 0.60*F_y*t*L_v]$
      $P_1 = 0.75*[58*0.25*3.50 + 0.60*36*0.25*6.75] = 65.4$ kips
   2. yielding on tensile plane and fracture on shear plane
      $P_2 = 0.75*[(\text{tensile yield force}) + (\text{shear fracture force})]$
      $P_2 = 0.75*[F_y*t*w + 0.60*F_u*t*L_v]$
      $P_2 = 0.75*[36*0.25*3.50 + 0.60*58*0.25*6.75] = 67.7$ kips

The design strength for block shear rupture is:
   $\phi P_n = 67.7$ kips (larger of P_1 and P_2)

For the tension member shown in Figure 3.5, the design strength is: $\phi P_n = 50.5$ kips (least of 50.5, 57.7, and 67.7 kips)

3.4 FILLET WELDS

Figure 3.6 shows possible cross sections of a fillet weld. The sides of the largest isosceles right triangle that can be inscribed within the cross section of the weld are called legs. The leg dimension is the fillet weld size. The effective throat thickness, t_e, is the shortest distance from the root of the weld to the hypotenuse of the isosceles right triangle. This definition of the effective throat thickness is for the SMAW (Shielded Metal Arc Welding) process. For the SAW (Submerged Arc Welding)

47

process there is a modification to the definition of the effective throat thickness(see LRFD J2.2 page 6-62).

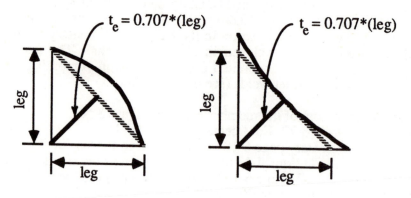

t_e is the *effective throat thickness*

Figure 3.6 Possible cross sections of a fillet weld

The design strength of a fillet weld is the smaller ϕR_n value obtained from:
1) shear or tension fracture on the weld throat plane
   [LRFD J2.2, J2.4 and Table J2.3 on LRFD page 6-65]

$$\phi R_n = 0.75*[0.60*F_{Exx}*(t_e*L_w)]$$

where leg = fillet weld size;
$t_e = 0.707*(leg)$
L_w = length of a weld
F_{Exx} = electrode yield strength

For example, for the longest fillet weld in Figure 3.5a,

$\phi R_n = 0.75*[0.60*70*(0.707*3/16)*6.75] = 28.2$ kips
2) shear fracture in the base material beneath a leg of the weld
   [LRFD J4 page 6-72]

$$\phi R_n = 0.75*[0.60*F_u*(leg)*L_v]$$

For example, for the longest fillet weld in Figure 3.5a,

$\phi R_n = 0.75*[0.60*58*(3/16)*6.75] = 33.0$ kips

Therefore, the design strength of the longest fillet weld in Figure 3.5a is 28.2 kips (smaller of 28.2 and 33.0 kips).

To find the design strength of the fillet weld group shown in Figure 3.5a, find the sum of the governing ϕR_n for each weld length:

$\phi R_n = 0.75*0.60*70*0.707*(3/16)*(1.85+3.50+6.75) = 50.5$ kips.

3.5 GUSSET PLATE WELDED TO MEMBER END

A *gusset plate* (a member end connector plate) generally has an irregular shape to accomodate the fastening of several member ends at a joint. The gusset plate shape in Figure 3.5a was chosen for simplicity in illustrating the following definitions.

The design strength of the gusset plate in Figure 3.5a is the least ϕR_n value obtained from:

1) yield strength in Section 3-3 [LRFD J5.2a page 6-73]

   $$\phi R_n = 0.9*[F_y*(w*t)]$$

   where w and t are the width and thickness of the gusset plate

   $\phi R_n = 0.9*[36*(5*5/16)] = 50.6$ kips

2) fracture strength in Section 3-3 [LRFD J5.2b page 6-73]

   $$\phi R_n = 0.75*F_u*A_n$$

   $\phi R_n = 0.75*58*5*(5/16) = 68.0$ kips

3) block shear rupture strength [LRFD J5.2c page 6-74]
   LRFD Eqn(J5-3) on page 6-74 is inconsistent with the block shear rupture strength equations given on LRFD page 6-187. The author recommends that the equations on LRFD page 6-187 be used and that LRFD Eqn(J5-3) be ignored until the inconsistency is resolved. On LRFD page 6-187 there are two block shear rupture figures for a gusset plate. Therefore, the author believes the LRFD Commentary writers intended to state that LRFD page 6-187 is also applicable for connecting elements. The larger value obtained from the equations on LRFD page 6-187 is the block shear rupture strength.
   a) tensile fracture and shear yielding (see Figure 3.7a)
   $$P_1 = 0.75*[F_u*(t*w) + 0.60*F_y*t*L_v]$$
   $P_1 = 0.75*[58*(5/16)*4.25 + 0.6*36*(5/16)*4.90] = 82.6$ kips
   b) tensile yielding and shear fracture (see Figure 3.7b)
   $$P_2 = 0.75*[F_y*(t*w) + 0.60*F_u*t*L_v]$$
   $P_2 = 0.75*[36*(5/16)*4.25 + 0.6*58*(5/16)*4.90] = 75.8$ kips

The block shear rupture design strength is: $\phi R_n = 82.6$ kips.
The design strength of the gusset plate is 50.6 kips (least of 50.6, 68.0, and 82.6 kips). For Figure 3.5, the design strengths obtained in Sections 3.3 through 3.5 were 50.5 kips for the member, 50.5 kips for the weld group, and 50.6 kips for the gusset plate. Therefore, in a factored load analysis the tension force in the member of Figure 3.5

cannot exceed 50.5 kips(least of 50.5, 50.5, and 50.6 kips) without violating the LRFD Specifications.

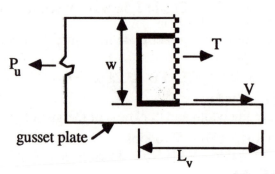

(a) Tensile fracture and shear yielding possibility

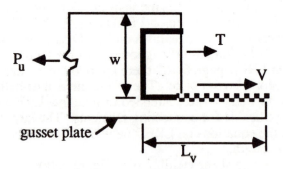

(b) Tensile yielding and shear fracture possibility

For the 5/16 by 5 gusset plate shown in Figure 3.5a:
   w = 3.50 + (5 - 3.50)/2 = 4.25 in.
   L_v = 6.75 - 1.85 = 4.90 in. .

Bold jagged lines indicate fracture planes.(▪▪▪▪▪▪)
Yield planes are shown as smooth surfaces.(————)

Figure 3.7   Block Shear Rupture possibilities for a gusset plate

3.6 DESIGN OF TENSION MEMBERS WITH WELDED CONNECTIONS

If there are not any holes in a tension member, for a design to be satisfactory the design strength must be equal to or exceed the required strength determined from a factored load analysis. Mathematically stated, this means $\phi P_n \geq P_u$ **is required** where ϕP_n is the design strength defined in Section 3.3 and P_u is the maximum tension force obtained from a factored load analysis. Therefore, if the member being designed is a single angle section fastened by fillet welds only on one leg of the angle to a gusset plate, the design requirements are:

1) for yielding in the gross section
   [LRFD D1 page 6-36]

   $(\phi P_n = 0.90 * F_y * A_g) \geq P_u$

2) for fracture in the member end
   [LRFD D1 page 6-36; LRFD B3 page 6-29]

   $[\phi P_n = 0.75 * F_u * (U * A_g)] \geq P_u$

3) (design strength for block shear rupture)
   [LRFD Commentary page 6-187]

The only unknown in each of the first two conditions is A_g. Since both conditions must be satisfied, the larger value of A_g obtained from these two conditions must be used in selecting the lightest available single angle section. After the single or double angle section with sufficient gross area is selected, the size, lengths, and locations of the fillet welds can be chosen. Then, the block shear rupture condition can be checked.

EXAMPLE 3.1 _____

See Figure 3.2 and consider the design of a tension member in the truss portion of this structure. Each truss member is to be a double angle section with long legs back to back and fillet welded to gusset plates at each truss joint. A detail of joint 12 is shown in Figure 3.8 to clarify the preceding description.

Member 34 in Figure 3.2 is to be designed using A36 steel (from LRFD page 6-123: $F_y = 36$ ksi; $F_u = 58$ ksi). $P_u = 71.9$ kips is the required strength obtained from Load Case 7 of the factored load analysis given in text Appendix A; assume the bending moment of 5.2 inch kips at joint 4 is negligible. Select the lightest available double angle section with long legs back to back that can be used. As shown in Figure 3.8, the long legs of each angle section are to be welded to a gusset plate.

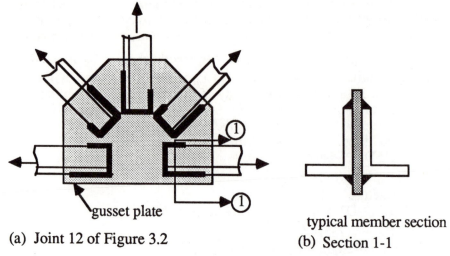

(a) Joint 12 of Figure 3.2

typical member section

(b) Section 1-1

Figure 3.8   Truss joint details for double angle members

SOLUTION--The design requirements which are a function of A_g are:
1) for yielding in the gross section
   $(0.90*F_y*A_g) \geq (P_u = 71.9$ kips)
   $A_g \geq [71.9/(0.90*36) = 2.22$ in^2]

2) for fracture in the member end
   $[0.75*F_u*(U*A_g)] \geq (P_u = 71.9$ kips)
   $U = 0.85$ (from item b on LRFD page 6-30)
   $A_g \geq [71.9/(0.75*58*0.85) = 1.94$ in^2]

From page 1-90 of LRFD Manual, select L3 X 2 X 1/4
$(A_g = 2.38$ in$^2) \geq 2.22$
weight = 8.2 lb/ft

Now that we have chosen the member, we can design the fillet welds and check the block shear rupture strength requirement. From LRFD page 6-63, the maximum size fillet weld that can be used in Figure 3.9 is (1/4) - (1/16) = 3/16 in. Use 3/16 in. E70xx fillet welds.

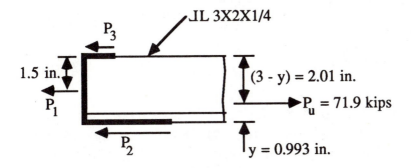

Figure 3.9 Fillet welds for the member chosen in Example 3.1

For a 1 in. weld length, the design strength is the smaller of:
1) shear or tensile fracture on the weld throat

$$\phi R_n = 0.75*[0.60*70*0.707*(3/16)*1] = 4.14 \text{ k/in}$$

2) shear fracture in the base material

$$\phi R_n = 0.75*[0.6*58*(3/16)*1] = 4.89 \text{ k/in}$$

The design requirement in Figure 3.9 is:
$(P_1 + P_2 + P_3) \geq (P_u = 71.9 \text{ kips})$ where:
$P_1 = 2*(4.14*3) = 24.8$ kips
$P_2 = 2*(4.14*L_2) = 8.28*L_2$
$P_3 = 2*(4.14* L_3) = 8.28*L_3$
Therefore, $8.28*(L_2 + L_3) \geq (71.9 - 24.8 = 47.1 \text{ kips})$ is required.
The author prefers to choose L_2 and L_3 such that the centroid of , P_1, P_2, and P_3 coincides with P_u. $\Sigma M_{P_3} = 0$ gives:

$3*P_2 + 1.5*P_1 = 2.01*P_u$
$P_2 = (2.01*71.9 - 1.5*24.8)/3 = 35.8$ kips
$8.28*L_2 = 35.8$
$L_2 = 4.32$ in.

Then, $8.28*L_3 \geq (47.1 - 35.8 = 11.3 \text{ kips})$
$L_3 \geq 1.36$ in.

From LRFD page 6-63, we find that the minimum acceptable weld length is 4 times the weld size and $4*(3/16) = 3/4$ in. The author chooses to use $L_2 = 4.50$ in. and $L_3 = 1.50$ in.

Now, we must check the requirement for the block shear rupture condition of the member end (see Figure 3.10). The design strength is the larger of:

1) tensile fracture and shear yielding

$P_1 = 2*0.75*[58*0.25*3.00 + 0.6*36*0.25*4.5] = 102$ kips

2) shear fracture and tensile yielding

$P_2 = 2*0.75*[0.6*58*0.25*4.5 + 36*0.25*3.00] = 99.2$ kips

Since ($\phi P_n = 102$ kips) $\geq$ ($P_u = 71.9$ kips), the design strength for block shear rupture is more than adequate.

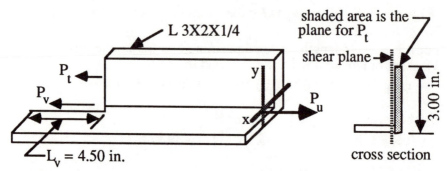

NOTE: member is a pair of angles; only one angle is shown above.

Figure 3.10 Block Shear Rupture model of member(Example 3.1)

EXAMPLE 3.2 _____

See Figure 3.2 and consider the design of a tension member in the truss portion of this structure. As shown in Figure 3.11, the top and bottom chord members of the truss are to be WT sections. Each truss web member is to be a double angle section with long legs back to back and fillet welded to the WT chord members at each truss joint. A detail of joint 12 is shown in Figure 3.11 to clarify the preceding description.

For simplicity in the discussion, assume that the same WT section is to be used in Figure 3.2 for members 5 through 14. From text Appendix A Load Case 7, we find that the required axial force variation(a minus sign is used in this text to denote a compression force) in the bottom chord is: -3.5 kips (member 5); 54.5 kips (member 6); 93.7 kips (member 7); 113.1 kips (member 8); and 120.8 kips (member 9). Also, we find that member 5 is predominantly a bending member with a bending moment of 88.4 inch kips at joint 2, but member 9 only has a bending moment of 21.2 inch kips at joint 10. How to cope with combined bending plus axial force is discussed in Chapter 6 of this text. For the purposes of the present example, assume that the required

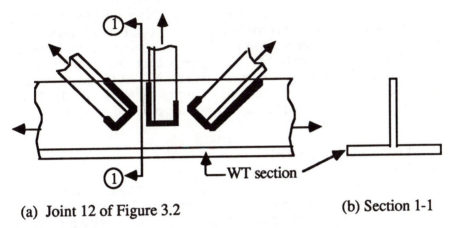

(a) Joint 12 of Figure 3.2   (b) Section 1-1

Figure 3.11 Truss joint details for WT bottom chord member

strength of members 5 through 14 is governed by members 9 and 10 for which the axial tension force is 120.8 kips. To compensate for the small bending moment, increase the axial force by 10%; thus, assume that P_u = 132.9 kips is the required strength for trial selection purposes. Use F_y = 36 ksi steel for which F_u = 58 ksi. Select the lightest available WT7 section that can be used for P_u = 132.9 kips.

SOLUTION--For trial selection purposes as a tension member, the design requirements which are a function of A_g are:
1) yielding in the gross section
$$(0.90*f_y*A_g) \geq (P_u = 132.9 \text{ kips})$$
$$A_g \geq [132.9/(0.90*36) = 4.10 \text{ in}^2]$$

2) fracture in the member end
$$[0.75*F_u*(U*A_g)] \geq (P_u = 132.9 \text{ kips})$$
$$A_g \geq [132.9/(0.75*58*1.00) = 3.06 \text{ in}^2]$$

U = 1.00 is assumed by the author to be appropriate in this design example for the following reasons: 1) members 5 through 14 are one continuous WT7 section; 2) the force transferred from the web members through their fillet welds to member 9 is only 120.8 - 113.1 = 7.7 kips which is negligible; and, 3) the force transferred from member 8 to member 9 does not come from any welds. Assuming U = 1.00 means we are saying that U is not applicable in this example. From LRFD page 1-70 , tentatively select WT 7 X 15 for which

$$(A_g = 4.42 \text{ in}^2) \geq 4.10 \text{ and weight} = 15 \text{ lb/ft.}$$

We will have to wait until the material in Chapter 6 of this text has been presented to check this trial selection for combined bending plus axial force. At that time we will be able to determine if the tentatively selected WT 7 X 15 can be used in Figure 3.2 for members 5 through 14. Since we do not know for certain that the WT7X15 is satisfactory, the author will not design the welds for this member and the block shear rupture condition requirement cannot be checked.

3.7 THREADED RODS

As shown in Figure 3.3, cross braces in roofs and walls may be designed as tension members to resist wind and to provide overall structural stability in a three dimensional sense for gravity type loads. If the roof slope in Figure 3.1 had been chosen to be greater than about 15 degrees, sag rods might be designed as tension members to provide lateral support for the weak axis of the purlins. In Figure 3.1, the sag rods would be perpendicular to the purlins and parallel to the roof. They would function in a manner similar to the saddle and stirrups for a horseback rider when the rider stands up in the stirrups. Each end of a sag rod would be threaded and passed through holes punched in each purlin web. A nut would be used on each end of a sag rod for anchorage. Adjacent sag rods would be offset in plan view about six inches or less to accomodate installation.

If rods are chosen as the tension members for cross bracing, a turnbuckle(see LRFD Manual page 5-174) may be used at midlength of the rod in order to take up slack or to pretension the rod. At each end of the rod, clevises(see LRFD Manual page 5-173) or welds may be used to fasten the rod to other structural members.

On LRFD page 6-36, the opening paragraph of Chapter D states that Sect. J3 is applicable for threaded rods. Table J3.2 on LRFD page 6-67 gives the tensile design strength of a threaded rod as:

$$\phi P_n = 0.75*(0.75F_u*A_g)$$

based on fracture through the threads. Table 1-B on LRFD page 5-3 can be used to select small diameter rods. Also, LRFD page 5-172 gives a table of gross areas and other data that can be used in the design of threaded rods.

EXAMPLE 3.3_____

Select threaded rods for the cross braces shown in Figure 3.3 using LRFD page 5-3 and A36 steel.

SOLUTION--Figure 3.1 gives the nominal wind load on the ends of the building as 12 psf pressure on the windward end and 7.5 psf suction on the leeward end. The total factored wind load on the building ends is $1.3*(0.012+0.0075)*(60*26.75) = 40.7$ kips.

Two pairs of cross braces are shown in Figure 3.3b. Hence, there are four identical pairs of cross braces that resist wind in the length direction of the building. At any time only one member in each pair of cross braces is in tension due to wind. The other member in each pair of cross braces is in compression due to wind and buckles at a negligibly small load. When the wind reverses the other member in each pair of cross braces resists wind. Thus, for each tension member the required strength is $P_u = 40.7*(32.41/20)/4 = 16.5$ kips. From LRFD page 5-3, select a 7/8 inch diameter threaded rod of A36 steel which has a design strength of $\phi P_n = 19.6$ kips.

3.8 TENSION MEMBER WITH BOLTED CONNECTIONS

In the fabrication shop, connections for building trusses usually are made by welding. If the entire truss is too large to ship, large segments of the truss are fabricated, shipped, and joined together by bolts on the job site to form the complete truss. Only bolted connections may be used for bridge trusses since the fatigue life of a bolted connection is greater than a welded connection.

Figure 3.12 shows the modes of failure for a bearing type bolted lap splice connection of a tension member. Separate design strength definitions are given for members, bolts, and connection plates. The governing design strength for a tension member with a bolted connection is the least of the design strength values for the: 1) member; 2) bolts; and, 3) connection plates. The definitions of design strength for bolts and connection plates are given later in this chapter.

57

Consider Figure 3.13 which shows a tension member fastened by bolts in a bearing type connection to a gusset plate. Along the member at Section 2 of Figure 3.13a which is some finite distance from the bolts, all cross sectional fibers can attain the yield strength provided the member end connection is stronger than the member. In the region of the member end connection, the stress distribution due to the applied load is not uniform in the member since some of the cross sectional elements are not bolted to the gusset plate. Hence, a transition region exists from the connection region to some finite distance from the connection where the stress distribution in the member becomes uniform when yield occurs. Thus, before yielding occurs in the member, the member end region containing the bolt holes usually experiences strain hardening and fracture can possibly occur in the member through the bolt holes.

If all cross sectional fibers of a member yield in tension, the member elongates excessively which can precipitate failure somewhere in the structural system of which the tension member is a part. Hence, tensile yielding of the gross section of the member is considered to be a limiting condition of failure. Fracture of the member can occur on Section 3 of Figure 3.13d which has a bolt hole in it. Also, a block shear fracture of the member can occur through all of the bolt holes on a section parallel to the member length(see Figure 3.14).

In the bearing type bolted connection of Figure 3.13a, each bolt is bearing on the member and on the gusset plate. Therefore, as shown in Figure 3.12d, a bearing failure can occur in the member. Bearing failure is analogous to laying a steel bolt across a stick of margarine. The bolt is harder than margarine. The bolt weight bears on and compresses the portion of the margarine stick beneath the bolt. Since bolts are much harder than the steel in the member and the gusset plate, almost all of the deformation due to bearing occurs in the member and gusset plate.

For simplicity purposes, the author chooses to give the mathematical definitions of the design strength for a tension member in EXAMPLE 3.4 and numerically illustrate the definitions for clarity.

EXAMPLE 3.4_____

The A36 steel member in Figure 3.13a is a single angle section, L 3.5 X 3 X 0.25, fastened by 3/4 inch diameter bolts in the long leg of the angle to a gusset plate. In Figure 3.13a, use a = 2 inches and C = 3 inches. Find the design strength of the member for a bearing type bolted connection.

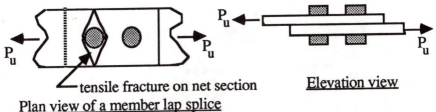

tensile fracture on net section   <u>Elevation view</u>

<u>Plan view of a member lap splice</u>

shaded items in each view are bolts in a bearing-type connection

(a) Tensile fracture on net section

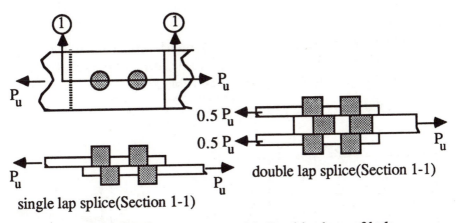

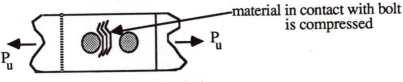

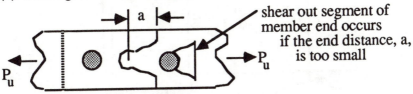

single lap splice(Section 1-1)   double lap splice(Section 1-1)

(b) Single shear of bolts   (c) Double shear of bolts

material in contact with bolt
is compressed

(d) Bearing failure caused by bolt

shear out segment of
member end occurs
if the end distance, a,
is too small

(e) Shear failure due to bearing

Figure 3.12 Bolted lap splice of a tension member

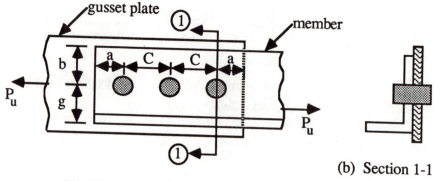

(b) Section 1-1

g = gage[LRFD page 5-166]
C = center to center distance between bolts[LRFD page 6-71]
a = end distance[LRFD page 6-72--Table J3.7 and Eqn(J3-3)]
b = edge distance[LRFD page 6-72--Table J3.7]

(a) Member end bolted to gusset plate

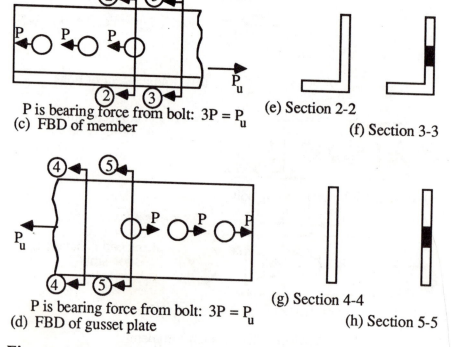

P is bearing force from bolt: $3P = P_u$ (e) Section 2-2
(c) FBD of member
                                                  (f) Section 3-3

(g) Section 4-4

P is bearing force from bolt: $3P = P_u$
(d) FBD of gusset plate   (h) Section 5-5

Figure 3.13 Tension member bolted to a gusset plate

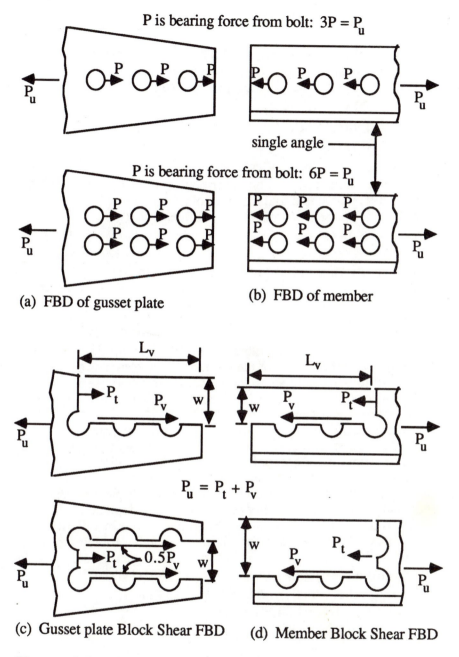

(a) FBD of gusset plate

(b) FBD of member

(c) Gusset plate Block Shear FBD

(d) Member Block Shear FBD

Figure 3.14 Examples of Block Shear failure

EXAMPLE 3.4 (continued)_____

SOLUTION--For a tension member with bearing type bolted connections the design strength is the least ϕP_n value obtained from:
1) yielding in the gross section(see Figure 3.13e)
   [LRFD D1 page 6-36]

   $\phi P_n = 0.90 * F_y * A_g$
   From LRFD page 1-56, $A_g = 1.56$

   $\phi P_n = 0.90*36*1.56 = 50.5$ kips
2) fracture in the member end on the critical section(see Figures 3.13c and 3.13f) which is through a bolt hole
   [LRFD D1 page 6-36] and [LRFD B3 page 6-30]

   $\phi P_n = 0.75 * F_u * A_e$
   where:
   A_e is the effective net area of the critical section
   $A_e = U * A_n$
   $A_n = A_g$ - (hole area in the cross section)
   hole area = (thickness of hole)*(diameter of hole)
   U is a reduction factor defined on LRFD page 6-30. U < 1 when the tension force in the member is transmitted at the member ends through some but not all of the cross sectional elements of the member to bolts and then to the gusset plate; otherwise, U = 1. For this example, U = 0.85 (see item b LRFD page 6-30).

It is implicitly understood that all bolt holes are standard holes except where the structural engineer specifies an oversized hole. In the fabrication shop, standard bolt holes are punched in the member provided the material thickness does not exceed the hole diameter. The nominal diameter of a standard bolt hole[LRFD J3 page 6-69] is the bolt diameter plus 1/16 inch. For tensile strength calculations, the hole diameter is defined[LRFD B2 page 6-29] as the nominal diameter of the hole plus 1/16 inch. Therefore, in this example problem, the hole diameter = (3/4 + 1/16) + 1/16 = 7/8 inch. In Figure 3.13f, the hole area = (hole diameter)*(thickness of hole) = (7/8)*(1/4) = 0.219 in^2

   $A_n = A_g$ - (hole area) = 1.56 - 0.219 = 1.34 in^2
   $A_e = U*A_n = 0.85*1.34 = 1.14$ in^2
   $\phi P_n = 0.75*(58*1.14) = 49.6$ kips

3) block shear rupture (see Figure 3.14d -- single angle FBD)
   [LRFD J4 page 6-72; Commentary page 6-186]
   According to the LRFD Commentary on LRFD page 6-188, the larger value obtained from the equations on LRFD page 6-187 is the block shear rupture strength. In Figure 3.14d, P_v is a shear force and P_t is a tensile force. Fracture can start to occur either on the shear plane or on the tension plane. When fracture starts to occur on one plane, the other plane will only be yielding.
   For shear calculations, the hole diameter is defined[LRFD B2 page 6-29] as the nominal diameter of the hole which is $3/4 + 1/16 = 13/16$ inch for this example.
   For Figure 3.13a, the member block shear FBD is as shown in Figure 3.14d. The length of the block shear plane is:
   (gross L_v) = a + 2C = 2 + 2*3 = 8 in.
   (net L_v) = 8 - 2.5*(13/16) = 5.97 in.
   A_{gt} = w*t = 1.5*0.25 = 0.375 in^2
   A_{nt} = A_{gt} - (hole area) = 0.375 - 0.219/2 = 0.266 in^2
   A_{gv} = (gross L_v)*t = 8*0.25 = 2.00 in^2
   A_{nv} = (net L_v)*t = 5.97*0.25 = 1.49 in^2

   $P_1 = \phi$*[(tensile fracture force) + (shear yield force)]
   P_1 = 0.75*[58*0.266 + 0.6*36*2.00] = 44.0 kips

   $P_2 = \phi$*[(shear fracture force) + (tensile yield force)]
   P_2 = 0.75*[0.6*58*1.49 + 36*0.375] = 49.0 kips

   ϕP_n = 49.0 kips (the larger of P_1 and P_2).

4) bearing strength at bolt holes
   [LRFD J3.6 page 6-68]
   [LRFD J3.9 page 6-71]
   [LRFD J3.10 page 6-72]
   In Figure 3.13a for this example problem, the end distance is a = 2 inches, the bolt spacing is C = 3 inches, and the bolt diameter is d = 0.75 inches.
   At an end edge bolt, if the end distance, a < 1.5d, the bearing failure mode shown in Figure 3.12e occurs and the bearing strength is given in LRFD J3.10. If a≥ 1.5d, the bearing strength at an end edge bolt is given in LRFD J3.6 where the end distance is denoted as L.
   For our example, (a = 2 in.) > (1.5d = 1.5*0.75 = 1.125 in.); therefore LRFD J3.10 is not applicable.

At an interior bolt, if the bolt spacing, $C \geq 3d$, the bearing strength is given in LRFD J3.6 and if $C < 3d$, the bearing strength is given in LRFD J3.9

For our example problem depicted in Figure 3.13a, there are three bolts in a line parallel tothe member tensile force and $(C = 3$ in.$) > (3d = 3*0.75 = 2.25$ in.$)$. Therefore, Eqn(J3-1a) on LRFD page 6-69 gives the applicable bearing strength at each interior bolt.

$\phi R_n = 0.75*(2.4*d*t*F_u) = 0.75*(2.4*0.75*0.25*58)$

$\phi R_n = 19.6$ k/bolt

For 3 bolts: $\phi R_n = 3*19.6 = 58.8$ kips

For the preceding example, it should be noted that:
1) the design aid shown on LRFD page 5-7 could have been used to obtain the design bearing strength = 19.6 k/bolt.
2) according to LRFD J3.10(Table J3.7) page 6-72, the minimum permissible value of the edge distance, b, in Figure 3.13a is $b = 1$ inch at a rolled edge. From LRFD page 5-166 the usual gage is $g = 2$ inches for an angle leg of 3.5 inches. Therefore, since the actual value of the edge distance is $b = 3.5 - 2 = 1.5$ is greater than the 1 inch minimum permissible value, the actual edge distance of 1.5 in. is okay.
3) the design aid on LRFD page 5-7 cannot be used to determine bearing strength at each end edge bolt if the end distance is less than 1.5 bolt diameters. LRFD J3.10 page 6-72 gives the design bearing strength at each end edge bolt if the end distance is less than 1.5 bolt diameters.

$$P = \phi*F_u*t*a$$

4) the design aid on LRFD page 5-7 cannot be used to determine bearing strength at each interior bolt if the bolt spacing is less than 3 bolt diameters. LRFD J3.9 page 6-71 gives the design bearing strength at each interior bolt if the bolt spacing is less than 3 bolt diameters.

$$P = \phi*[F_u*t*(C - (d + 1/16)/2)]$$

Suppose the minimum permissible bolt spacing(see LRFD page 6-71), $C = 2.67*d = 2.67*0.75 = 2.00$ in., is used instead of the 3 in. bolt spacing in the author's example. Then, the design bearing strength at each interior bolt is:

$P = 0.75*[58*0.25*(2-(13/16)/2)] = 17.3$ k/bolt

Summary for the originally posed Example 3.4:
Of the four computed values, the least value was 49.0 kips (due to block shear fracture). Therefore, the design strength of the tension member is: $\phi P_n = 49.0$ kips.

3.9 BOLTS IN BEARING TYPE CONNECTION

As shown in Figures 3.12b and 3.12c, failure of a bolt in a bearing type bolted connection is due to shear and each bolt can be subjected either to single shear or to double shear. LRFD J3.3 page 6-66 gives the shear design strength of a bolt as:

ϕR_n = 0.65*[(nominal strength)*(unthreaded cross sectional area)]

The nominal strength(units are: ksi) is obtained from LRFD Table J3.2 page 6-67. For a bolt subjected to single shear, the unthreaded cross sectional area of one bolt is appropriate. For a bolt subjected to double shear, there are two shear failure planes. Therefore, the shear design strength for a bolt subjected to double shear is two times the shear design strength of a bolt subjected to single shear.

EXAMPLE   3.5_____

Find the shear design strength of the bolt group shown in Figure 3.13a assuming 3/4 in. diameter A325 bolts are used and the bolt threads are excluded from the single shear plane.

SOLUTION--From LRFD page 6-67, the shear nominal strength is 72.0 ksi and the cross sectional area of a circular area having a diameter of 3/4 in. is 0.4418 in^2. $\phi R_n = 0.65*(72*0.4418) = 20.7$ k/bolt

For the 3 bolt group, $\phi R_n = 3*20.7 = 62.1$ kips.

It should be noted that the author could have used the design aid on LRFD page 5-5 to find the *single shear design strength* = 20.7 k/bolt for one A325X 3/4 in. diameter bolt. See the footnotes on LRFD page 5-5 for the definitions of connection types N and X and the loading types S and D. The title of LRFD page 5-5 should have been written as follows: "Shear Design Strength in kips".

3.10 GUSSET PLATE BOLTED TO MEMBER END

As shown in Figure 3.8a, more than one member is usually fastened to an irregularly shaped gusset plate. Some, but not necessarily all, of the attached members may be in tension. In the following discussion we are only concerned with the design strength of the gusset plate at the end of each attached tension member. The simplest possible case is one tension member attached to a gusset plate. Equally simple is a lap splice for a tension member(see Figure 3.12c). Essentially, a lap splice plate and the simplest case of a gusset plate are very short tension members. Therefore, except for the definition of fracture on the critical net section, the tensile design strength definitions of a lap splice and the simplest case of a gusset plate are identical to the definitions of the design strength for a tension member. For simplicity purposes, the author chooses to give the mathematical definitions of the design strength for a gusset plate in an example problem where the definitions can be numerically illustrated for clarity.

EXAMPLE 3.6_____

In the bearing type bolted connection of Figure 3.13a, the A36 steel gusset plate is 5 in. wide by 5/16 in. thick and has 3/4 inch diameter bolts in it. In Figure 3.13a, use a = 2 inches and C = 3 inches. Find the design strength of the gusset plate.

SOLUTION--For a gusset plate in a bearing type bolted connection, the design strength is the least ϕR_n value obtained from:
1) yielding in the gross section(see Figure 3.13g)
   [LRFD J5.2 page 6-73]

   $\phi P_n = 0.90*F_y*A_g$
   $A_g = 5*(5/16) = 1.56$ in^2
   $\phi R_n = 0.9*36*1.56 = 50.6$ kips

2) fracture in the critical net section(see Figure 3.13h)
   [LRFD J5.2 page 6-73]
   [LRFD B2 page 6-29 if there are staggered holes]
   Our gusset plate is in tension, but is very short compared to the length of an attached tension member. The behavior of the gusset plate is not identical to a tension member. Therefore, the following LRFD restriction applies for our gusset plate and for a

66

tension member splice plate. If the **computed A_n** > 0.85*A_g, the definition of A_n = 0.85*A_g

$$\phi R_n = 0.75*F_u*A_n$$

$A_n = A_g$ - (hole area in the cross section)
hole area = (thickness of hole)*(diameter of hole)
    Each bolt hole diameter = (3/4 + 1/16) + 1/16 = 7/8 inch for tension calculation purposes. In Figure 3.13f, the hole area = (diameter)*(thickness) = (7/8)*(5/16) = 0.273 in^2
A_g = 5*(5/16) = 1.56 in^2
0.85*A_g = 0.85*1.56 = 1.33 in^2
(A_n = 1.29 in^2) < (0.85 = 1.33); therefore, A_n = 1.29 in^2
ϕR_n = 0.75*58*1.29 = 56.1 kips

3) block shear rupture (see Figure 3.14c -- one row of bolts FBD case)
    [LRFD J5.2c page 6-74]
    For shear calculations, hole diameter = 3/4 + 1/16 = 13/16 in.
    (gross L_v) = a + 2C = 2 + 2*3 = 8 in.
    (net L_v) = 8 - 2.5*(13/16) = 5.97 in.
    A_{gt} = w*t = 2.5*(5/16) = 0.781 sq. in.
    A_{nt} = 0.781 - 0.273/2 = 0.645 in^2
    A_{gv} = (gross L_v)*t = 8*5/16 = 2.5 in^2
    A_{nv} = (net L_v)*t = 5.97*(5/16) = 1.87 sq. in.
    P_1 = 0.75*[(tensile fracture force) + (shear yield force)]
    P_1 = 0.75*(58*0.645 + 0.6*36*2.50) = 68.6 kips
    P_2 = 0.75*[(tensile yield force) + (shear fracture force)]
    P_2 = 0.75*(36*0.781 + 0.6*58*1.87) = 69.9 kips

    ϕP_n = 69.9 kips (the larger of P_1 and P_2).

4) bearing strength at bolt holes
    [LRFD J3.6 page 6-68]
        Bolt diameter, d = 0.75 in. In Figure 3.13a, the end distance is a = 2 inches and the bolt spacing is C = 3 inches. LRFD Spec. J3.10 is not applicable since
                (L = a = 2 in.) > (1.5d = 1.5*0.75 = 1.125 in.).
        Since (C = 3 in.) > (3d = 3*0.75 = 2.25 in.), LRFD J3.9 is not applicable. There are 3 bolts in a line parallel to the member tensile force. Therefore, Eqn(J3-1a) on LRFD page 6-69 gives the applicable bearing strength for each bolt. (Also, the design aid on LRFD page 5-7 is applicable.)

ϕR_n = 0.75*(2.4*d*t*F_u) = 0.75*[2.4*0.75*(5/16)*58]

$\phi R_n = 24.5$ k/bolt

For 3 bolts: $\phi R_n = 3*24.5 = 73.5$ kips

Summary for Example 3.6:
Of the four strength values, the least value was 50.6 kips (due to yielding in the gross section). Therefore, the gusset plate design strength is: $\phi R_n = 50.6$ kips.

Summary of the tension member with a bearing type bolted connection:

1) member(Example 3.4): $\phi P_n = 49.0$ kips due to block shear rupture

2) bolt group shear(Example 3.5): $\phi R_n = 62.1$ kips

3) gusset plate(Example 3.6): $\phi R_n = 50.6$ kips due to yielding in the gross section

The least of these three values is the governing design strength. According to the LRFD Specifications, the design is satisfactory provided the required strength does not exceed the governing design strength. For the case being discussed, the required strength is the maximum tension force in the member due to a factored load analysis.

3.11 EFFECT OF STAGGERED HOLES ON NET AREA

Figure 3.15a is the FBD of a tension member separated from a bearing type bolted connection. The governing design strength of the member is the least value due to: 1) bearing; 2) tensile yielding in the gross section; 3) tensile fracture in the critical net section; and, 4) block shear fracture. Suppose we have determined that the governing design strength is due to tensile fracture in the critical net section. The net section in Figure 3.15c has the largest internal force and all of the net sections have the same net area. Therefore, Section 1-1 is the critical net section since fracture would occur in this section if the member were loaded to failure.

In Figure 3.16a the bolt holes are staggered to provide a larger net area than the bolt configuration in Figure 3.15a would give. Our discussion will be limited to the determination of the design strength due to tensile fracture in the critical net section. If fracture occurs in the net section shown in Figures 3.16c and 3.16d, the fracture lies in one plane. However, if fracture occurs in the net section of Figures 3.16e and 3.16f, the fracture does not lie in one plane. Figures 3.16d and 3.16h have the same net area and the same fracture design strength.

68

However, the internal force due to equilibrium requirements is smaller in Figure 3.16h than in Figure 3.16d. Furthermore, the internal force in all net sections on the left side of Figure 3.16h is not greater than the internal force in Figure 3.16h. Therefore, the critical net section(where the fracture would occur) is either Figure 3.16c or Figure 3.16e. The design strength due to tensile fracture in the net section is the smaller value obtained for the net sections in Figures 3.16c and 3.16e.

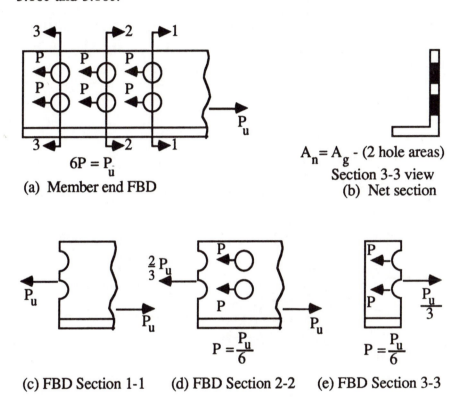

(a) Member end FBD

$6P = P_u$

$A_n = A_g$ - (2 hole areas)
Section 3-3 view
(b) Net section

(c) FBD Section 1-1

(d) FBD Section 2-2

$P = \dfrac{P_u}{6}$

(e) FBD Section 3-3

$P = \dfrac{P_u}{6}$

Figure 3.15 Tension member with two lines of bolts

In the calculations of the design strength due to tensile fracture of the member in Figure 3.16a:
one hole area = (diameter)*(thickness) = 0.875*0.5 = 0.438 in². For Figure 3.16c, $A_n = A_g$ - (one hole area) = 4.50 - 0.438 = 4.06 in². Since one leg of the angle section member does not contain any bolts

69

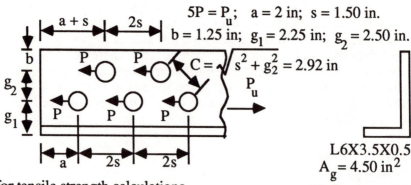

$5P = P_u$; $a = 2$ in; $s = 1.50$ in.
$b = 1.25$ in; $g_1 = 2.25$ in; $g_2 = 2.50$ in.
$C = \sqrt{s^2 + g_2^2} = 2.92$ in

L6X3.5X0.5
$A_g = 4.50$ in^2

(b) Gross section

for tensile strength calculations,
hole diameter = 0.875 inches

(a) Member end FBD

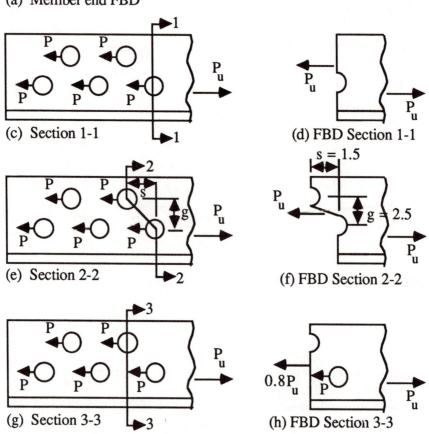

(c) Section 1-1

(d) FBD Section 1-1

(e) Section 2-2

(f) FBD Section 2-2

(g) Section 3-3

(h) FBD Section 3-3

Figure 3.16 Staggered bolt configuration in a tension member

and the top line contains only two bolts, U = 0.75 (see item c on LRFD page 6-30). For Figure 3.16d,

$\phi P_n = 0.75*(F_u*A_e) = 0.75*[F_u*(U*A_n)] = 0.75*58*(0.75*4.06)$
$\phi P_n = 132$ kips.

LRFD B2 page 6-29 gives an empirical definition of the net area as:

$$A_n = A_g - A_{holes} + \sum \frac{s^2 t}{4g}$$

which is applicable for Figures 3.16e,f:
$A_n = 4.50 - 2*0.438 + (1.5)^2*0.5/(4*2.5) = 3.74$ in^2
For Figure 3.16f, $\phi P_n = 0.75*58*(0.75*3.74) = 122$ kips.

    For the member in Figure 3.16a, the design strength due to tensile fracture is 122 kips (smaller of 132 kips and 122 kips). If the governing design strength for this member is due to tensile fracture in the critical net section, the fracture would occur in the net section of Figure 3.16c.

    In Figure 3.17a, all cross sectional elements of the member contain bolts; therefore, U = 1 is applicable in the effective net area definition. For vizualization purposes in the net area calculations, flatten out the angle as shown in Figure 3.17c.

    For the net section that passes through only hole B:
$A_n = 4.50 - 0.438 = 4.06$ in^2
$[\phi P_n = 0.75*58*4.06 = 177$ kips$] = T$
    For the net section that passes through holes A and C:
$A_n = 4.50 - 2*0.438 = 3.62$ in^2
$[\phi P_n = 0.75*58*3.62 = 157$ kips$] = [T - P = (6/7)*T]$
$T = (7/6)*157 = 183$ kips would be required to produce tensile fracture on this net section.
    For the net section that passes through holes A and B:
$A_n = 4.50 - 2*0.438 + (1.5)^2*0.5/(4*2.5) = 3.74$ in^2
$[\phi P_n = 0.75*58*3.74 = 163$ kips$] = T$
    For the net section that passes through holes A, B, and C (see Figure 3.17c):
$A_n = A_g - A_{holes} + \sum[s^2*t/(4*g)]$
$A_n = 4.50 - 3*0.438 + (1.5)^2*0.5/(4*2.5) + (1.5)^2*0.5/(4*3.75)$
$A_n = 3.37$ in^2
$[\phi P_n = 0.75*58*3.37 = 147$ kips$] = T$

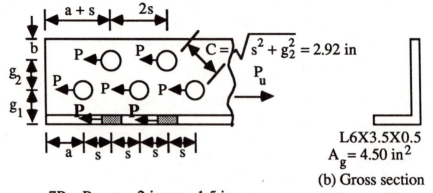

$C = \sqrt{s^2 + g_2^2} = 2.92$ in

L6X3.5X0.5
$A_g = 4.50$ in^2

(b) Gross section

$7P = P_u$; a = 2 in; s = 1.5 in.

b = 1.25 in; $g_1 = 2.25$ in; $g_2 = 2.50$ in.;

for tensile strength calculations, hole diameter = 0.875 inches

(a) Member end FBD

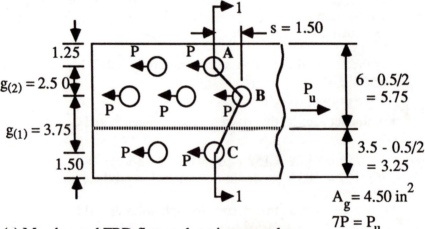

(c) Member end FBD flattened out into one plane

Figure 3.17 Bolts in all cross sectional elements

The critical net section is the net section that requires the least value of T to produce tensile fracture. Since the least T = 147 kips was for the net section shown in Figure 3.17c, the tensile fracture design strength for the member in Figure 3.17a is: $\phi P_n = 0.75*58*3.37 = 147$ kips.

3.12 DESIGN OF TENSION MEMBERS WITH BOLTED CONNECTIONS

The design strength must be equal to or exceed the required strength determined from a factored load analysis. Therefore, the design requirements are:

1) for yielding in the gross section

$$(\phi P_n = 0.90*F_y*A_g) \geq P_u$$

2) for fracture in the critical net section

$$[\phi P_n = 0.75*F_u*(U*A_n)] \geq P_u$$

3) **(sum of bearing design strength at the bolt holes) $\geq P_u$**
4) **(block shear rupture design strength) $\geq P_u$**

EXAMPLE 3.7_____

Member 34 in Figure 3.2 is to be designed using A36 steel. From Load Case 7 in text Appendix A, $P_u = 71.9$ kips is the required strength. Select the lightest available pair of angles with long legs back to back that can be used in a bearing type bolted connection. One line of 3/4 in. diameter A325N bolts in double shear will be placed in the long legs of the angles. Use: 3 in. bolt spacing; 1.5 in. end distance; and standard size bolt holes.

SOLUTION--From LRFD page 5-5, $\phi R_n = 31$ k/bolt for double shear. The minimum number of bolts is (71.9 kips)/(31 k/bolt) = 2.3 bolts. Use 3 bolts.

LRFD page 5-7 can be used to determine the minimum thickness needed for each angle at each bolt hole since (a = 1.5 in.) $\geq$ (1.5d = 1.5*0.75 = 1.125) and (C = 3 in.) $\geq$ (3d = 3*0.75 = 2.25). Each of the 3 bolts bears on the long leg of each angle. Therefore, the bearing design strength needed at each bolt hole in each angle is:

$\phi R_n \geq [71.9/(2*3) = 12.0$ kips].

$F_y = 58$ ksi for A36 steel(see LRFD page 6-123). On LRFD page 5-7 for t = 3/16 in., ($\phi R_n = 14.7$ kips) $\geq$ 12.0 kips. Therefore, the minimum thickness of each angle is 3/16 in. for the bearing design strength requirement to be satisfied.

For yielding in the gross section, the design requirement is:

$$(\phi P_n = 0.90*F_y*A_g) \geq (P_u = 71.9 \text{ kips})$$

$A_g \geq [71.9/(0.90*36) = 2.22$ in^2] is required.

73

For fracture in the critical net section, the design requirement is:

$[\phi P_n = 0.75*F_u*(U*A_n)] \geq (P_u = 71.9$ kips)

$U = 0.85$ is obtained from item b on LRFD page 6-30 since there are 3 bolts per line.

$A_n \geq [71.9/(0.75*58*0.85) = 1.94$ in^2] is required.

One bolt hole area $= d*t = (0.75 + 0.125)*t = 0.875*t$

The critical net section has a bolt hole in each long leg of each angle; therefore, $A_n = A_g - 2*(0.875*t) = A_g - 1.75*t$.

Summary of the design requirements:
1. $t \geq 3/16$ in. for bearing
2. $A_g \geq 2.22$ sq. in. for yielding in the gross section
3. $[A_n = A_g - 1.75*t] \geq 1.94$ in^2 for fracture in the net section
   See LRFD page 1-90:
   a) for $t = 3/16$ in. $= 0.1875$ in.
      need: $A_g \geq (1.94 + 1.75*0.1875 = 2.27$ in$^2) \geq 2.22$ in^2
      No available section for $t = 3/16$ in. has $A_g \geq 2.27$ in^2.
   b) for $t = 1/4$ in. $= 0.25$ in.
      need: $A_g \geq (1.94 + 1.75*0.25 = 2.38$ in^2) ≥ 2.22 in^2
      choose: L3X2X1/4 ($A_g = 2.38$) $\geq (2.38 \geq 2.22)$
   c) for $t = 5/16$ in. $= 0.3125$ in.
      need: $A_g \geq (1.94 + 1.75*0.313 = 2.49$ in^2) $\geq (2.22$ in^2)
      choose: L2.5X2X5/16 ($A_g = 2.62$in^2) $\geq (2.49 \geq 2.22)$
   d) for $t = 3/8$ in. and so on
      Proceed as shown in items 1 and 2 until it becomes obvious that the lightest available choice has been found. In this example the only reason for doing the $t = 5/16$ in. case was to illustrate the general procedure. Since the pair of angles chosen for the $t = 1/4$ in. case has exactly the minimum required gross area, no lighter choice is possible for any other acceptable thickness.

Check the block shear rupture strength requirement for the lightest available acceptable pair of angles which is L3X2 X1/4

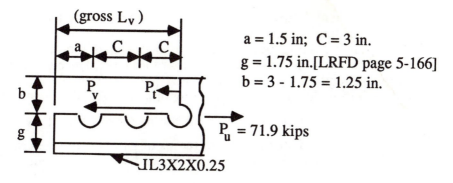

a = 1.5 in; C = 3 in.
g = 1.75 in.[LRFD page 5-166]
b = 3 - 1.75 = 1.25 in.

P_u = 71.9 kips

⌐L3X2X0.25

Figure 3.18 Block Shear Rupture of member end

(gross L_v) = a + 2C = 1.5 + 2*3 = 7.5 in.
(net L_v) = 7.5 - 2.5*(0.75 + 0.0625) = 5.47 in.
A_{gv} = 2*(7.5*0.25) = 3.75 in^2
A_{nv} = 2*(5.47*0.25) = 2.74 in^2
A_{gt} = 2*(1.25*0.25) = 0.625 in^2
A_{nt} = 2*(1.25*0.25 - 0.875*0.25/2) = 0.406 in^2
P_1 = 0.75*(58*0.406 + 0.6*36*3.75) = 78.4 kips
P_2 = 0.75*(0.6*58*2.74 + 36 *0.625) = 88.4 kips

Since (ϕP_n = 88.4 kips) ≥ (P_u = 71.9 kips), the block shear rupture
design strength requirement is satisfied.
Use pair of L3X2X1/4   weight = 8.2 lb/ft

3.13   STIFFNESS CONSIDERATIONS

After the structure in Figure 3.2 is erected, from the factored load
combinations in text Appendix A we find that member 34 is required to
resist a maximum axial force of 71.9 kips in tension and 2.6 kips in
compression. In the fabrication shop and in shipping the prefabricated
truss, a crane is used to lift the prefabricated truss. Unless special
lifting procedures are used, member 34 may be required to resist a
larger compressive axial force due to lifting than is required after the
structure is erected.

The design strength of a compression member is given in Chapter
4 of this text. If the axial compressive force in a member becomes too
large, the member buckles and may precipitate collapse of the entire
structure. For a pinned ended member the buckled shape is a half sine

wave along the length direction of the member. For clarity purposes, this type of buckling is called column buckling.

If column buckling occurs for a pinned ended member, the member bends about the cross sectional axis of least resistance which is the minor principal axis. A fundamental parameter in the design strength definition due to column buckling of a pinned ended member is the maximum slenderness ratio, $\frac{L}{r}$, of the member where L is the member length and r is the minimum radius of gyration for the cross section of the member. As shown in text Appendix B, the definition of radius of gyration for any cross sectional axis is $r = \sqrt{\frac{I}{A}}$ for that axis where I is the moment of inertia and A is the gross area. The minimum moment of inertia is found from the minor principal axis and is used to compute the minimum radius of gyration needed to obtain the maximum slenderness ratio. The design requirement is that column buckling must not occur before the required axial compressive strength is developed in the member.

LRFD B7 page 6-33 states that preferably $\frac{L}{r}$ should not exceed 300 for a tension member except for threaded rods. The reason(see LRFD page 6-149) for this advisory upper limit on $\frac{L}{r}$ is to provide adequate bending stiffness for ease of handling during fabrication, shipping, and erection. If the tension member will be exposed to wind or perhaps subjected to mechanically induced vibrations, a smaller upper limit on $\frac{L}{r}$ may be needed to prevent excessive vibrations.

EXAMPLE 3.8

For the tension member designed in Example 3.1, compute the maximum slenderness ratio.

SOLUTION--From Example 3.1 we find that:
1) member 34 of Figure 3.2 was designed; L = 90 in.
2) the design choice was a pair of L3X2X1/4 with the long legs separated by and welded to a gusset plate at each member end. For the purposes of this example, assume the gusset plate hickness is 3/8 in.

76

LRFD page 1-90 contains a sketch of the cross section of the member and the radius of gyration for each principal axis. As shown in text Appendix B, if the cross section has an axis of symmetry then the axis of symmetry is a principal axis and the other principal axis is perpendicular to the axis of symmetry. On LRFD page 1-90, since the axis of symmetry is the Y-axis, the principal axes are the X and Y axes. Therefore, for member behavior as a pair of L3X2X1/4 separated 3/8 in. back to back, the minimum r is the smaller of:

$$r_x = 0.957 \text{ in. and } r_y = 0.891 \text{ in.}$$

For behavior as a pair of angles:

the maximum $\dfrac{L}{r}$ = (90 in.)/(0.891 in.) = 101 which is less than the preferred upper limit of 300.

It should be noted that in the figure on LRFD page 1-90 the gusset plate exists only at the member ends. Elsewhere along the member length, there is a 3/8 in. gap between the long legs of the pair of angles. Double angle behavior is truly ensured only at the points where the pair of angles is tied together(by the gusset plates at the member ends). Therefore, we need to determine the individual behavior of each angle. See LRFD page 1-59. For one L3X2X1/4, the minimum r is $r_z = 0.435$ in. and the maximum L/r = 90/0.435 = 207 which is less than the preferred upper limit of 300.

If we insert a 3/8 in. plate between the long legs at the midlength of the member and weld the long legs to this plate, we will ensure double angle behavior at midlength of the member as well as at the member ends. Then, the length for single angle behavior is L = 90/2 = 45 in. and the maximum L/r = 45/0.435 = 103.5 for single angle behavior. Since 103.5 is very nearly equal to 101(for double angle behavior), we can conclude that we only need to insert and weld a spacer plate at midlength of the member in order to ensure that double angle behavior is valid.

PROBLEMS

NOTATION

P_u is the required axial strength of a tension member

ϕP_n is the design tensile strength of a member
   [see LRFD p 6-36 and 6-30 if welded connections are used]
   [also see p 6-29 if bolted connections are used]

ϕR_n is the design shear rupture strength
   [see LRFD p 6-72,73,187]

See LRFD p 6-64,65 for the design strength of fillet welds.

See LRFD p 6-66 to 6-72 for the design strength of bolted connections.
   Use the tables on LRFD p 5-3 to 5-7 when they contain the needed
   design strength information.

NOTE to students:

Your professor is to specify the grade of steel and weld electrodes to be
used in problems 3.1 to 3.5. If your professor forgets to specify these
parameters, **assume: A36 steel and E70 electrodes.**

3.1 7/16 inch fillet welds configured as shown in Figure P3.1

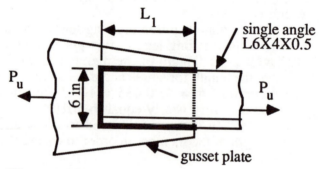

Figure P3.1

Assume that the gusset plate will be adequately designed later.

a) Find ϕP_n for yielding in the gross section

b) Find ϕP_n for fracture in the member end

c) For P_u = [smaller ϕP_n found in (a) and (b)], find the minimum
   acceptable value of L_1 to satisfy the design strength
   requirement for the welds.

d) Using L_1 found in (b), find the block shear rupture strength, ϕR_n

e) What is the maximum acceptable value of P_u?

3.2 5/16 inch fillet welds if 36 ksi steel is specified
3/8 inch fillet welds if 50 ksi steel is specified
7/16 inch fillet welds if 65 or 100 ksi steel is specified

Find the maximum acceptable value of P_u for the truss joint detail shown in Figure P3.2.

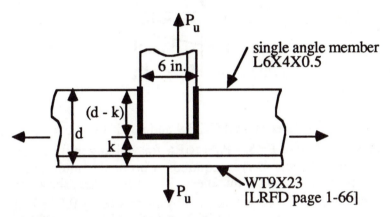

Figure P3.2 Truss joint detail

3.3 $P_u = 270$ kips

Select the lightest acceptable pair of angles with a 3/4 inch separation to serve as a tension member. Using the maximum acceptable weld size and the weld configuration shown in the sketch for problem 3.1, design the welds assuming that the block shear design strength does not govern the minimum value of L_1.

Find ϕR_n. If $\phi R_n < 270$ kips, find the minimum value of L_1 such that $\phi R_n = 270$ kips.

3.4 $P_u = 300$ kips and 5/8 inch fillet welds. Select the lightest acceptable channel to serve as a tension member. The channel ends are connected by fillet welds to an adequately designed gusset

plate. Ignore block shear design strength and find the minimum
acceptable value of L_1.

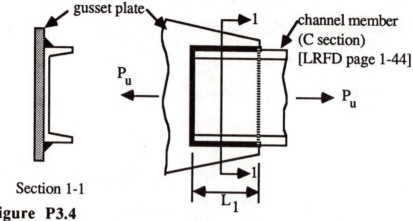

Section 1-1

Figure P3.4

3.5 A tension member is a C15X33.9 and needs to be butt spliced as
shown in Figure P3.5. $P_u = (\phi P_n$ for yielding in the gross
section of the member) is required. Choose the minimum
acceptable splice plate thickness. The splice plate, shown cross
hatched in the cross section, is t by 14 by $(2L_1 + 0.5$ in.). Fillet
welds are located as shown in the cross section and on the back
side of the web of the C15X33.9 at the ends of the splice plate.
Find the minimum acceptable value of L_1 assuming that block
shear does not govern. Check block shear of the splice plate and
of the member.

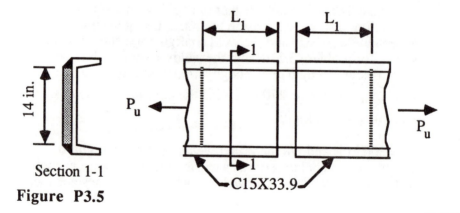

Section 1-1

Figure P3.5

NOTE to students:

Unless your professor specifies otherwise, a bearing-type bolted connection and punched holes are to be used in solving the following problems. The usual gages on LRFD page 5-166 are to be used for angles. Your professor is to specify the grade of steel and bolts, the bolt diameter, and s to be used. If your professor forgets to specify these parameters, assume:

A36 steel; 1 inch diameter A325N bolts;

C = larger of (3 bolt diameters and 3 in.).

Assume the gusset plate does not govern P_u.

3.6 Find the maximum acceptable value of P_u.

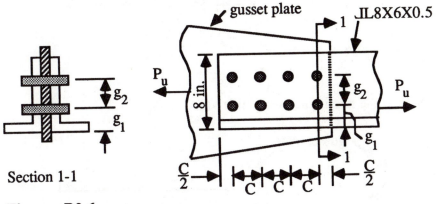

Section 1-1

Figure P3.6

3.7 Find the maximum acceptable value of P_u.

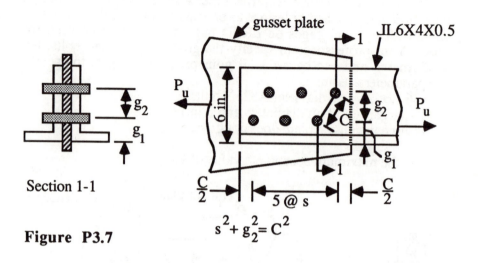

Section 1-1

Figure P3.7

$$s^2 + g_2^2 = C^2$$

3.8 Find the maximum acceptable value of P_u.

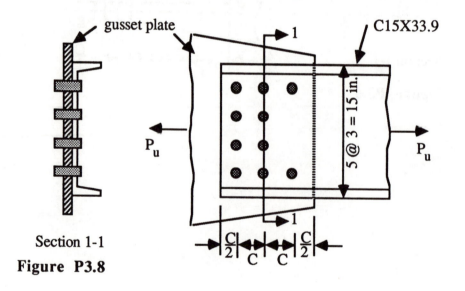

Section 1-1

Figure P3.8

82

3.9 Each splice plate is 0.375 by 10 by [6X + 0.5 in.]
The member is a 0.75 by 10 plate.
Find the maximum acceptable value of P_u.

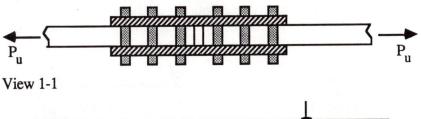

View 1-1

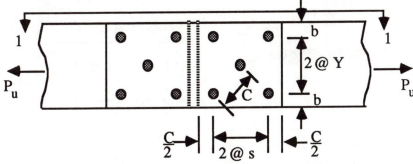

b = minimum rolled edge distance[LRFD page 6-72]

s = 0.707*C

Y = (10 -2b)/2

Figure   P3.9

3.10 A pair of 6X4X1/2 angles is used as a tension member. The member ends are bolted to gusset plates. All cross sectional elements of the member have bolts in them. The bolt holes are staggered and are located as shown in the figure below which shows one angle flattened out into a plane. Assume the gusset plates do not govern and find the maximum acceptable value of P_u.

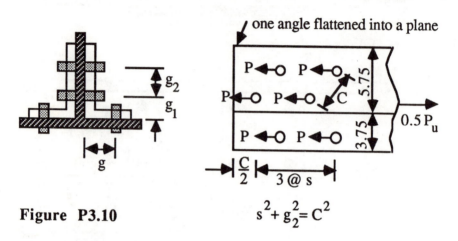

one angle flattened into a plane

Figure P3.10

$$s^2 + g_2^2 = C^2$$

3.11 P_u = 400 kips
For eight A325N bolts located as shown in the sketch for problem 3.6, find the minimum bolt diameter required and the lightest acceptable pair of angles.

3.12 P_u = 300 kips
For ten A325N bolts located as shown in the sketch for problem 3.8, find the minimum bolt diameter required and the lightest acceptable C section.
Use C = 3.75 inches.

Chapter 4

COMPRESSION MEMBERS

4.1 INTRODUCTION

A member subjected only to an axial compressive force is called a *column*. Practically speaking, it is impossible for a member to be subjected only to an axial compressive force. When a lab test of such a member is conducted, locating the centroid of the member's cross section at the member ends in order to apply the axial force concentrically cannot be done perfectly. Also, the member is not perfectly straight. Consequently, the initial crookedness increases as the axial compression force is applied. Hence, the member is subjected to a combination of bending and axial compression; such a member is called a *beam-column* (see Chapter 6). However, as we will see in Chapter 6, the design strength definition of a *beam-column* is an interaction equation that contains a term due only to column action and another term due only to bending action. This chapter deals with column action assuming that bending is negligible or will be accounted for later.

In a truss analysis, only an axial tension or an axial compression force is assumed to exist in each member. Also, there are other situations where the designer deems that bending is negligible and considers only *column action* in the design of a compression member. Consequently, for practical reasons as well as for the subsequent treatment of *beam-columns*, we need to consider *column action* as a separate topic.

Consider Figure 4.1 which shows a compression member bolted to a gusset plate. Examination of the FBD in Figure 4.1d shows that the maximum compressive force in the net section for a bearing type bolted connection is always less than the axial compressive force in the gross section. Therefore, if the only holes in a compression member are at the member ends for a bolted connection, the net section is not involved in the design strength definition of a column.

When the axial compressive force in a pinned ended member, for example, reaches a certain value called the *critical load* or *buckling load*, the member buckles. For a pinned ended member whose cross section is doubly symmetric as shown in the figure at the top of LRFD page 1-36, the *buckled shape* is a half sine wave in the X-X plane along the length direction of the member. This type of buckling is called *column buckling*. If the flange and web elements of the cross

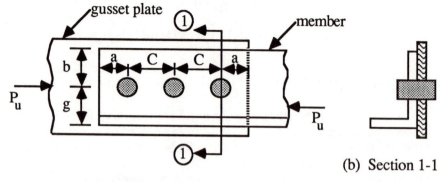

g = gage[LRFD page 5-166]
C = spacing of bolts[LRFD page 6-71]
a = end distance[LRFD page 6-72--Table J3.7 and Eqn(J3-3)]
b = edge distance[LRFD page 6-72--Table J3.7]

(a) Member end bolted to gusset plate

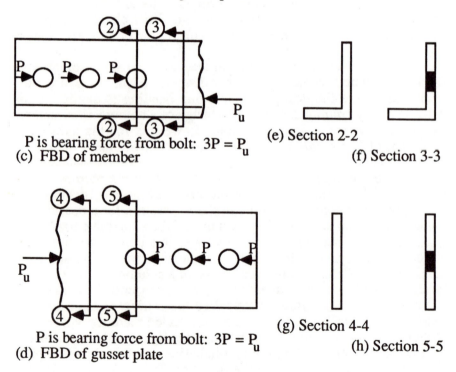

P is bearing force from bolt: $3P = P_u$
(c) FBD of member

(e) Section 2-2

(f) Section 3-3

P is bearing force from bolt: $3P = P_u$
(d) FBD of gusset plate

(g) Section 4-4

(h) Section 5-5

Figure 4.1 Compression member bolted to a gusset plate

section on LRFD page 1-36 are not properly configured, the author will show in Section 4.7 that *local buckling of these elements* can occur before column buckling occurs. Therefore, both types of buckling must be discussed in this chapter. For the W sections usually chosen to serve as a column, the flange and web elements are properly configured to prevent local buckling from occurring before column buckling occurs. Consequently, column buckling is discussed first in this chapter. For the simplest type of column buckling, no twisting of the cross section occurs when the column buckles. If the cross section twists when column buckling occurs, the discussion in text Section 4.6 is applicable.

All of the example problems in this chapter are shown in Section 4.8 after all aspects of column behavior have been discussed since the column design strength definition may have to account for local buckling of the cross sectional elements.

4.2 ELASTIC EULER BUCKLING OF COLUMNS

As shown on LRFD page 6-151, the boundary conditions of a column significantly influence the buckled shape which can be used to compute the buckling load. The first theoretical buckling load solution was published in 1744 by Leonhard Euler[4] for a flagpole column, Case (e) on LRFD page 6-151. In 1759 he[5] published the solution for a pinned ended column, Case (d) on LRFD page 6-151. In his solutions, Euler assumed the member was elastic and perfectly straight before the axial compressive load was applied. As the axial load was slowly applied, he assumed the member remained elastic and perfectly straight until the value of the axial load reached the *critical load* or *buckling load*. Then, he reasoned that the member had reached a state of critical equilibrium and buckled into an assumed shape which was dependent on the boundary conditions at the member ends. Consequently, as Euler chose to do, we choose to call the value of the axial compressive load at which the member buckles as the *critical load*.

The prismatic, pinned ended member in Figure 4.2 is assumed to be perfectly straight before the axial compressive load, P, is slowly applied. Also, the member is assumed to be weightless and elastic. When P reaches the critical load value, P_{cr} , the member bows into a bent configuration and the cross section does not twist when bowing occurs. Bowing or bending occurs about the cross sectional axis of least resistance which is the minor principal axis, v. Note that the member deflects in the direction of the major principal axis, u, which is

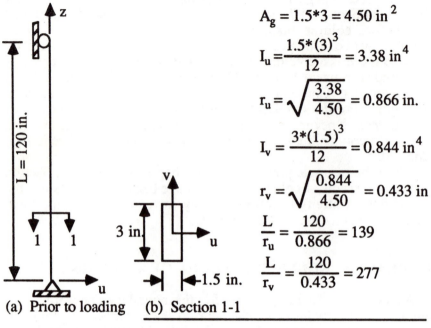

$$A_g = 1.5*3 = 4.50 \text{ in}^2$$

$$I_u = \frac{1.5*(3)^3}{12} = 3.38 \text{ in}^4$$

$$r_u = \sqrt{\frac{3.38}{4.50}} = 0.866 \text{ in.}$$

$$I_v = \frac{3*(1.5)^3}{12} = 0.844 \text{ in}^4$$

$$r_v = \sqrt{\frac{0.844}{4.50}} = 0.433 \text{ in}$$

$$\frac{L}{r_u} = \frac{120}{0.866} = 139$$

$$\frac{L}{r_v} = \frac{120}{0.433} = 277$$

(a) Prior to loading   (b) Section 1-1

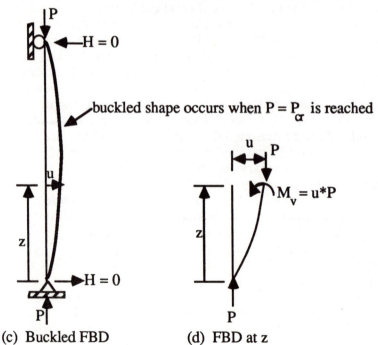

buckled shape occurs when $P = P_{cr}$ is reached

$M_v = u*P$

(c) Buckled FBD   (d) FBD at z

Figure 4.2 Elastic pinned ended column

perpendicular to the bending axis. At an arbitrary distance, z, from the origin of the member, the bending moment in the buckled member is:

$$M_v = u*P \qquad \text{(Eqn 4.1)}$$

which can be substituted into the moment-curvature relation:

$$\frac{d^2u}{dz^2} = \left(\frac{M_v}{EI_v} = \frac{P*u}{EI_v}\right) \qquad \text{(Eqn 4.2)}$$

to obtain the governing differential equation:

$$\frac{d^2u}{dz^2} + \frac{P}{EI_v}u = 0 \qquad \text{(Eqn 4.3)}$$

For mathematical convenience, let $c^2 = \dfrac{P}{EI_v}$; then Eqn(4.3) becomes:

$$\frac{d^2u}{dz^2} + c^2u = 0 \qquad \text{(Eqn 4.4)}$$

The solution of Eqn(4.4) is:

$$u = A \sin cz + B \cos cz \qquad \text{(Eqn 4.5)}$$

The second derivative of Eqn(4.5) with respect to z is $-c^2*u$. If we substitute this second derivative value into Eqn(4.4), we find that indeed Eqn(4.5) is the solution for Eqn(4.4).

At $z = 0$, the boundary condition is $(u = 0) = A \sin 0 + B \cos 0$ which requires $B = 0$.

At $z = L$, the boundary condition is $(u = 0) = A \sin cL$ which can be satisfied in any of the following ways:
1) $A = 0$
   If we accept this possibility, $A = 0$ and $B = 0$ substituted into Eqn(4.5) gives $u = 0$ everywhere along the member which means that the member remains straight(does not deflect). Since the member bows into a deflected shape when P reaches the critical load value, we must reject this possibility.

89

2) $\sin cL = 0$

   If we choose $c = 0$, then $\sin 0 = 0$ and $c^2 = 0 = \dfrac{P}{EI_v}$ which requires $P = 0$ since EI_v is not zero. This possibility, $c = 0$, must be rejected since $P = 0$ means no load was applied.

3) $\sin cL = 0$

   If we choose $cL = \pi$, then $\sin \pi = 0$ and $(cL)^2 = \pi^2$ which gives:

$$\left(c^2 = \frac{P}{EI_v}\right) = \frac{\pi^2}{L^2} \qquad \text{(Eqn 4.6)}$$

Therefore, the critical load or buckling load for Figure 4.2 is:

$$P_{cr} = \frac{\pi^2 EI_v}{L^2} \qquad \text{(Eqn 4.7)}$$

Substituting $I_v = A_g = r_v^2$ into Eqn(4.7) and then dividing both sides of the resulting equation by A_g gives the *critical stress*:

$$F_{cr} = \frac{P_{cr}}{A_g} = \frac{\pi^2 E}{\left(\dfrac{L}{r_v}\right)^2} \qquad \text{(Eqn 4.8)}$$

It should be noted that we could not find a value for the coefficient A in Eqn(4.5). We only determined that A must be greater than zero. Therefore, the buckled shape in Figure 4.2 is a half sine wave with an indeterminate amplitude, $A > 0$.

4.3 EFFECT OF INITIAL CROOKEDNESS ON COLUMN BUCKLING

Since perfectly straight members cannot be manufactured, each rolled steel section has an initial curvature upon arrival at the fabrication shop. If the *out-of-straightness*, $e \leq \dfrac{L}{1000}$ (see LRFD page 6-254) the member will not be straightened(some of the crookedness will not be removed). Prior to loading of a pinned ended column, e is the maximum deflection relative to a straight line connecting the member ends and L is the member length. For a rolled steel section, the <u>average value of e</u> $= \dfrac{L}{1500}$ (see LRFD page 6-155).

The behavior of an initially crooked member subjected to axial compression is shown in Figure 4.3 where the maximum out- of-straightness was assumed to occur at midheight of the member. L/r is a fundamental parameter in Eqn(4.8) which is the critical stress definition for an initially straight, pinned ended member. $\frac{L}{r}$ is called the slenderness ratio; L is the member length and r is the radius of gyration for the axis about which bending would occur for Euler buckling(initially perfectly straight member). The effect of initial crookedness on the critical load is greatest in the range $50 < \frac{L}{r} < 135$ for a column of A36 steel. Based on experimental and theoretical investigations in which the cross section did not twist when buckling occurred, the LRFD Specification writers chose Eqn(4.9) as the critical load definition of an elastic, prismatic, pinned ended column with an out-of-straightness of $e = \frac{L}{1500}$.

$$P_{cr} = 0.877*\frac{\pi^2 EI_v}{L^2} \qquad \text{(Eqn 4.9)}$$

4.4   INELASTIC BUCKLING OF COLUMNS

As explained in Section 2.4, the residual stress pattern shown in Figure 4.4a is representative of some W sections. The *maximum compressive residual stress*, f_{rc}, occurs at the flange tips and at midheight of the web; and, the *maximum tensile residual stress*, f_{rt}, occurs at the junction of the flanges and the web.

Consider an axial compression laboratory test of a pinned ended W section member of A36 steel. Assuming the member is perfectly straight and twisting does not occur during buckling, elastic buckling occurs provided (see Figures 4.4a and 4.4b) the maximum compressive residual stress plus the applied stress when buckling occurs is less than F_y. For A36 steel, if f_{rc} + (Eqn 4.8) < 36 ksi, none of the compression fibers are yielding and elastic buckling occurs(see Figure 4.4c). If we choose $\frac{L}{r}$ small enough to prevent elastic buckling from occurring, the compressive stress-strain curve(see Figure 4.4c) for a W section containing residual stresses becomes nonlinear after the flange tips begin to yield and inelastic buckling occurs. As shown in Figure 4.4c, the slope of the compressive stress-strain curve for inelastic buckling is called the *tangent modulus of elasticity*, E_t.

91

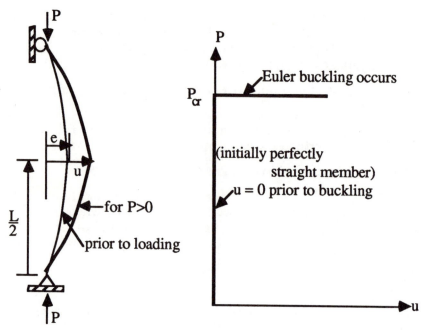

(a) Initially crooked member   (b) Load vs Deflection for e = 0

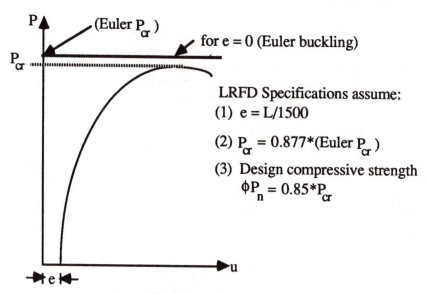

(c) Load vs Deflection for initially crooked member

Figure 4.3 Initially crooked, elastic, pinned ended column

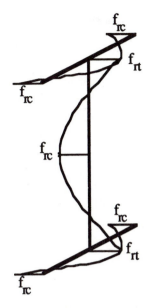

(a) Residual stresses

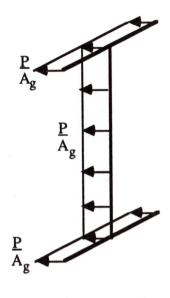

(b) Axial compression stress

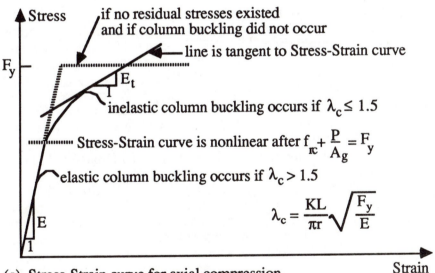

(c) Stress-Strain curve for axial compression

Figure 4.4 Perfectly straight W section member

93

If twisting does not occur during buckling of a perfectly straight pinned ended column, the inelastic critical load is:

$$P_{cr} = \frac{\pi^2 E_t I_v}{L^2}$$

(Eqn 4.10)

If the column has an out-of-straightness of $e = \dfrac{L}{1500}$, the LRFD inelastic critical load definition is:

$$P_{cr} = 0.877 * \frac{\pi^2 E_t I_v}{L^2}$$

(Eqn 4.11)

and the corresponding inelastic critical stress is:

$$F_{cr} = 0.877 * \frac{\pi^2 E_t}{\left(\dfrac{L}{r_v}\right)^2}$$

(Eqn 4.12)

After the author has discussed the *effective length concept* in text Section 4.5, the definition of E_t will be derived.

4.5   EFFECTIVE LENGTH

The *effective length* (**equivalent pinned ended length**), KL, of a column is the distance between the points of inflection (M = 0 points) on the buckled column shape. See LRFD page 6-151 for the K values of some buckled shape examples for isolated, individual columns. L is the actual length of each column and KL is the <u>chord distance between the M = 0 points on each buckled shape</u>. KL must be obtained for each principal axis of the cross section. For example, an individual W section column may be fixed at the base and free at the top (see Case (e) on LRFD page 6-151) for bending about the major principal axis, but may be hinged at the top (see Case (b) on LRFD page 6-151) for bending about the minor principal axis.

For a column in an **unbraced frame**, see Figure C-C2.1 on LRFD page 6-151, the nomograph on LRFD page 6-153 can be used to obtain a very good approximate value of K for in-plane, elastic column buckling. For a column in a **braced frame**, the *sidesway inhibited* nomograph on LRFD page 2-5 can be used to obtain K. After the author has derived an expression for the tangent modulus, E_t, the definition for G on LRFD pages 2-5 and 6-153 will be revised to account for inelastic column buckling and for more general girder end boundary conditions.

94

In LRFD E2 page 6-39[see LRFD Eqn(E2-2)], the inelastic critical stress definition is:

$$F_{cr} = (0.658^{\lambda_c^2})*F_y \qquad \text{(Eqn 4.13)}$$

where

$$\lambda_c = \frac{KL}{\pi r}\sqrt{\frac{F_y}{E}} \qquad \text{(Eqn 4.14)}$$

Eqn(4.14) must be evaluated for each principal axis and the larger value obtained for Eqn(4.14) must be used in Eqn(4.13) to obtain the critical stress. **KL/r** is called the *slenderness ratio* and L is the distance between braced points for each principal axis.

The following definition based on LRFD critical stress definitions is obtained by equating Eqns(4.12) and (4.13) and solving for E_t/E:

$$\frac{E_t}{E} = \frac{\lambda_c^2(0.658^{\lambda_c^2})}{0.877} \qquad \text{(Eqn 4.15)}$$

See Figure 4.5 which summarizes the LRFD definition of the critical load for a prismatic, axially loaded, compression member that: does not twist when column buckling occurs by bending about the principal axis with the larger KL/r ratio; contains a realistic, representative, residual stress pattern; and, has an out-of-straightness of e = L/1500.

The theoretical bases for the nomographs on LRFD pages 6-153 and 2-5 need to be examined and revised to account for inelastic column buckling and various boundary conditions at the far ends of the girders. In an unpublished paper, Julian and Lawrence[11] made the assumptions stated on LRFD page 6-152 in deriving the equations for the effective length factor, K, for a column in a plane frame. They used one of the derived equations in preparing the nomograph on LRFD page 6-153 for an unbraced frame. The other derived equation was used in preparing the nomograph on LRFD page 2-5 for a braced frame(*sidesway inhibited*).

In preparing the nomograph on LRFD page 6-153, all members were assumed to be elastic and a point of inflection (M = 0) was assumed to occur at midspan of each girder in the frame. A girder (restraining member) is a bending member that is attached to one or more column ends at a particular joint in a frame. Each girder provides a rotational resistance at the column end(s) when column buckling

95

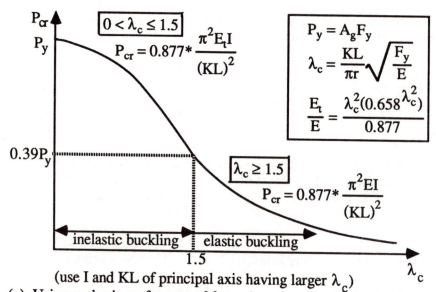

(use I and KL of principal axis having larger λ_c)

(a) Using author's preference of formulas: P_{cr} is as defined above

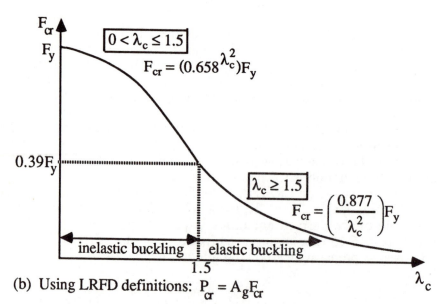

(b) Using LRFD definitions: $P_{cr} = A_g F_{cr}$

NOTE: In (a) and (b), λ_c must be evaluated for each principal axis and the larger λ_c must be used.

Figure 4.5 Column Design Strength, $\phi P_{cr} = 0.85 * P_{cr}$

buckling occurs for the frame. It must be noted that an important assumption which was not listed on LRFD page 6-152 is: the effect of any axial compression force in each girder was ignored by the nomograph authors in their girder end rotational stiffness definition. Therefore, at each end of a girder, the girder end rotational stiffness was assumed to be $\frac{3EI}{0.5L} = \frac{6EI}{L}$. To account for inelastic column buckling and the $M = 0$ point not being at midspan of the elastic girders in using the nomograph on LRFD page 6-153 to estimate K for a column in an unbraced frame, the definition of the *relative joint stiffness parameter*, **G**, is:

$$G = \frac{\sum \left(\frac{E_t}{E} \frac{I}{L}\right)_c}{\sum \left(\gamma \frac{I}{L}\right)_g} \qquad \text{(Eqn 4.16)}$$

where E_t/E is as defined in Eqn(4.15) and

$$\gamma = \frac{\textbf{actual girder end rotational stiffness}}{6EI/L} \qquad \text{(Eqn 4.17)}$$

For example: $\gamma = \frac{3EI/L}{6EI/L} = 0.5$ at joint B in Figure 4.6

$\gamma = \frac{0}{6EI/L} = 0$ at joint C in Figure 4.6

$\gamma = \frac{4EI/L}{6EI/L} = 2/3$ at joint B in Figure 4.7

$\gamma = \frac{6EI/L}{6EI/L} = 1$ at each girder end in Figure 4.8

In preparing the **braced frame**(*sidesway inhibited*) **nomograph** on LRFD page 2-5, all members were assumed to be elastic and each girder in the braced frame was assumed to bend in single curvature such that the girder end rotations were identical. The effect of any axial compression force in a girder was ignored by the nomograph authors in their girder end rotational stiffness definition. Therefore, at each end of a girder, the girder end rotational stiffness was assumed to be $\frac{2EI}{L}$. To account for inelastic column buckling and unequal girder end rotational stiffnesses for the elastic girders in using the sidesway inhibited

97

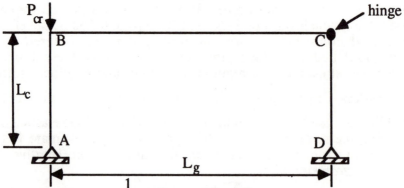

Figure 4.6 $\gamma = \frac{1}{2}$ at B; $\gamma = 0$ at C

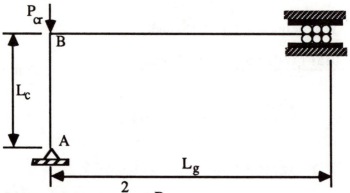

Figure 4.7 $\gamma = \frac{2}{3}$ at B

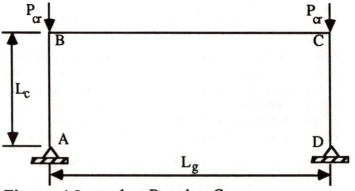

Figure 4.8 $\gamma = 1$ at B and at C

nomograph on LRFD page 2-5 to estimate K for a braced frame, the definition of the relative joint stiffness parameter, G, is Eqn(4.16) where $\frac{E_t}{E}$ is as defined in Eqn(4.15) and

$$\gamma = \frac{\textbf{actual girder end rotational stiffness}}{2EI/L} \qquad \text{(Eqn 4.18)}$$

For example, in Figure 4.9:

$$\gamma = \frac{3EI/L}{2EI/L} = 1.5 \quad \text{at joint A for girder 1}$$

$$\gamma = \frac{4EI/L}{2EI/L} = 2 \quad \text{at joint C for girder 3}$$

$$\gamma = \frac{2EI/L}{2EI/L} = 1 \quad \text{at each end of girder 2}$$

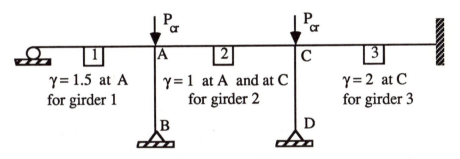

$\gamma = 1.5$ at A for girder 1 $\gamma = 1$ at A and at C for girder 2 $\gamma = 2$ at C for girder 3

Figure 4.9 Braced frame

4.6 FLEXURAL-TORSIONAL BUCKLING OF COLUMNS

In all of the previous discussions in this chapter, it was assumed that the buckled shape of a column was due to bending about the principal axis with the larger KL/r value and that the cross section did not twist when column buckling occurred. This is called the *flexural mode of column buckling*. However, it is possible that a doubly symmetric cross section(W section, for example) only twists when the column buckles; this is called the **torsional mode** of column buckling. Singly symmetric cross sections(T section, channel, equal

99

legged angle) in which the shear center does not coincide with the centroid (see Fig. 4.10), and unsymmetric cross sections(an angle with unequal legs and built-up sections) bend and twist when the column buckles; this is called the *flexural-torsional mode of column buckling*. Theoretical discussions of *torsional* and *flexural-torsional modes* of elastic column buckling are available[6,7,8].

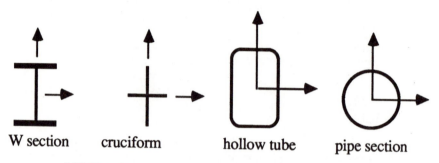

W section   cruciform   hollow tube   pipe section

NOTE: Shear center coincides with centroid.

(a) Doubly symmetric sections

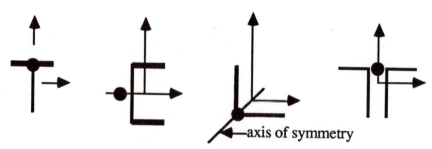

—axis of symmetry

NOTE: Dot shows location of shear center.
          Shear center does not coincide with centroid.

(b) Singly symmetric sections

Figure 4.10   Column cross sections

For unsymmetric cross sections, the *critical load* must be determined from the flexural-torsional column buckling mode. See LRFD E3 pages 6-39 and 6-90.

If the end supports and any intermediate weak axis column braces are properly designed to prevent twisting of a W section, only *flexural column buckling* can occur **provided local buckling is prevented**. However, if the intermediate weak axis column braces are designed to prevent only a translation perpendicular to the weak axis and do not prevent twisting of a W section, the unbraced length for torsion is the member length whereas the maximum distance between the intermediate braces is the unbraced length for flexure. Consequently, if the intermediate weak axis column braces do not prevent twisting of a W section, the *critical load* is the smaller value obtained from the *torsional column buckling mode* [see LRFD Eqn(A-E3-5) page 6-91] and the *flexural column buckling mode*.

For built-up cross sections, see LRFD E4 page 6-40. Suppose the member is a double angle section(see sketch on LRFD page 2-66) separated by and connected to a 3/8 inch thick gusset plate at each member end. In order to ensure double angle member behavior, the two angles must be connected to each other by placing a 3/8 inch thick spacer plate between them at one or more intermediate locations along the member length. Each spacer plate is fastened to the two angles by welding or by bolting. The spacer plate location(s) must be such that L_s/r for the minor principal axis of one angle is not greater than the maximum KL/r value for the double angle behavior. For single angle behavior, L_s is the maximum length between two adjacent spacer (connector) plates. However, L is the member length for double angle behavior.

If a compression member in a truss is a single angle section with only one angle leg fastened at the member ends to a gusset plate, the compression force is applied eccentrically loaded as shown on LRFD page 2-49. The member is subjected to biaxial bending plus compression and should be treated as a *beam-column* as illustrated in Example 9 on LRFD page 2-49.

If a compression member in a truss is a double angle section with only one leg of each angle fastened at the member ends to a gusset plate, the modified KL/r definition on LRFD page 6-40 is applicable for the flexural mode of column buckling and item b on LRFD page 6-91 is applicable for the *flexural-torsional mode of column buckling*. The author gives an example in text Section 4.8 to illustrate the design procedure for such a member.

4.7 LOCAL BUCKLING OF THE CROSS SECTIONAL ELEMENTS

Most likely each reader either has made or has seen someone make a paper airplane by folding a sheet of notebook paper into a desired airplane shape. The author assumes that the airplane maker folded up a small strip at the wing tips to stiffen the wing tips in regards to bending. This strip is called an *edge stiffener* or a lip(a projecting edge -- Webster's dictionary definition). One edge of a lip projects into the air, is not fastened to anything, and is called a free edge.) The interior edges of the wings were stiffened by the fuselage. Therefore, along the length direction of the airplane, each wing edge was stiffened and we can classify each of the described paper airplane wings as a *stiffened element*. Each lip is called an *unstiffened* (**projecting**) *element* since one lip edge is free and the other lip edge is attached to another element.

It should not be difficult for the reader to visualize the essential parts of the following discussion. Suppose the described paper airplane is set on its tail end with the nose pointing upward. A small compression force can be applied in the gravity direction at the nose of the airplane before buckling is observed. The buckled shape is not easily vizualized and the author cannot describe the buckled shape since the airplane configuration varies depending on each airplane maker's creative ability. If a flat sheet of the same notebook paper could be stood on one end, at best the flat sheet of notebook paper would barely support its own weight before buckling was observed. The author's point is that an identical amount of material can be made into various shapes and some shapes are better for resisting a compression load.

Steel pipes(LRFD page 2-34) and structural steel tubes(LRFD page 2-37) are very good shapes for a column cross section, but attaching other members to these shapes can be difficult and expensive. Some examples of *stiffened* and *unstiffened* compression elements in column cross sectional shapes are shown in Figure 4.11 to further clarify the definitions for compression elements.

Examples of the *local buckling mode shapes* for the flange and for the web of a W section are shown in Figure 4.12 assuming that local buckling occurs. Also, in Figure 4.12 the author assumed that the flange and the web buckled independently of each other. Theoretical discussions of *elastic buckling of thin plate elements* are available[see Chapter 9 of references 6 and 7 which show that Eqn(4.19) is the **critical stress** for elastic buckling of thin plate elements).

$$F_{cr} = \frac{k\pi^2 E}{12(1-\upsilon^2)*(b/t)^2} \qquad \text{(Eqn 4.19)}$$

where **k** is a constant depending on L/t (length/thickness ratio) and the boundary conditions of the plate element edges parallel to the direction of applied compressive stress. In Eqn(4.19), the other parameters are:

E, modulus of elasticity; υ, Poisson's ratio; and **b/t,** width/thickness ratio of the plate element.

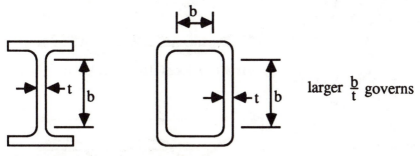

(a) Stiffened compression elements

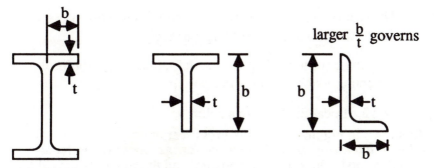

(b) Unstiffened(projecting) compression elements

Figure 4.11 Compression elements in column cross sections

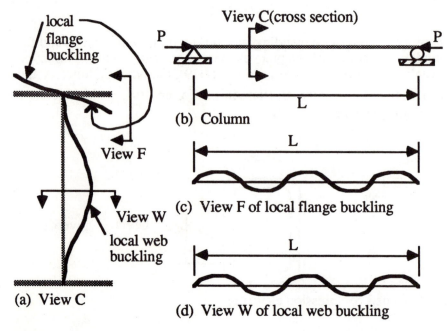

(a) View C

View C(cross section)

(b) Column

(c) View F of local flange buckling

(d) View W of local web buckling

NOTE: Number of half sine waves in (c) and (d) depends on L and b/t of flange and b/t of web, respectively.

LRFD Appendix B[page 6-87] gives the definition of F_{cr} for local buckling of compression elements in a column.

Figure 4.12 Local Buckling in a column

As shown in Figure 4.12 at any cross section:
1) *if local buckling of a W section flange occurs*, the local buckling mode shape is antisymmetric and the web only provides a small amount of rotational resistance which can be ignored. Parallel to the applied compressive stress, one flange edge is free and the other edge(at the junction of the flange and web) can be conservatively assumed to be a hinge. **Each half of a W section flange** is an *unstiffened compression element.*

2) *if local buckling of a W section web occurs*, the local buckling mode is symmetric and the flanges provide considerable rotational resistance since for *column buckling*, the structural designer must prevent twisting of the cross section at the member ends and at any intermediate, weak axis, column braced points. Therefore, the flanges are restrained at the ends of each unbraced

104

column length and the torsional resistance of the flanges can be developed in each unbraced column length. At each junction of the web and flanges, the web edge is somewhere between fully fixed and hinged. The **web** is a *stiffened compression element.*

It should be noted that:
1) *if local buckling of the compression elements in a column cross section occurs*, these elements continue to resist some more compressive load until a considerable amplitude of the buckled shape occurs. However, the reader should recall that *if column buckling occurs* the member cannot resist any more compressive axial load.
2) **inelastic** *local buckling of plates can occur* if either the b/t ratio or the L/t ratio is small enough.

For routine situations, the reader fortunately does not have to deal with Eqn(4.19). Realistic values of **k** for Eqn(4.19) were chosen by the AISC Specification writers(see LRFD page 6-4) based on currently available theoretical and experimental research, satisfactory performance of existing structures, and engineering judgment to devise definitions of critical stress for local buckling of column cross sectional elements.

If the b/t ratio of each compression element in a column cross section is less than the limiting value in LRFD Table B5.1 page 6-32, local buckling does not occur before column buckling occurs and the design strength of a column is given by LRFD E2 page 6-39.

If local buckling of a compression element in a column cross section occurs before column buckling occurs, LRFD B3.5 page 6-33 refers the reader to LRFD Appendix B5.3 page 6-87 for the design strength definition of a column. The effective portion of each stiffened and unstiffened compression element in a column cross section must be determined and used in the LRFD Appendix B definition of the design strength for a column.

The author does not permit his students to use LRFD Appendix B in the first steel design course. Each column section must be selected from only those sections for which local buckling does not occur before column buckling occurs. However, the students are required to show that the b/t ratio of each compression element in the selected column section does not exceed the limiting value given in LRFD Table B5.1 page 6-32.

4.8 EXAMPLE PROBLEMS

First, throughout the sequence of the example problems, LRFD Specifications and design aids are used to find the axial compression design strength of a column. These kind of problems are called investigation or checking problems since everything is known except for the design strength. Then, examples of how to select a column section or how to design a column are presented.

In order to enable the reader to quickly locate an example for a specific type of problem, an index summary description of the example problems is shown below.

Example	Brief Description
4.1	Find W section column design strength using LRFD formulas.
4.2	EX 4.1 with different KL values for each principal axis.
4.3	EX 4.1 solved using LRFD Column Design Strength Table.
4.4	EX 4.2 solved using LRFD Column Design Strength Table.
4.5	Design example. Select W14, W12, W10, W8 columns.
4.6	Unbraced frame nomograph on LRFD page 6-153 is used to find the effective length of some columns.
4.7	Design example. EX 3 on LRFD page 2-6 is correctly done.
4.8,9	EX 4.7 is repeated for a different girder size.
4.10	Find column design strength for a double angle section.
4.11	Design example. Select double angle section.

EXAMPLE 4.1_____

Given:

An A36 steel W14X90 section is to be used as an axially loaded compression member and the effective length is KL = 20 ft for both principal axes.

Prove that local buckling does not occur before column buckling occurs. Then, use the formulas in Figure 4.5 to find the axial compression design strength.

SOLUTION See LRFD page 1-30 for the following W14X90 properties which are in inch units:

A = 26.5; d = 14.02; k = 1.375

web: $\dfrac{h_c}{t_w} = 25.9$ **flange:** $\dfrac{0.5b_f}{t_f} = 10.2$

An axis of symmetry is a principal axis; therefore, the X and Y axes are the principal axes. Since the X axis has the larger moment of inertia value, the X axis is the major principal axis and the Y axis is the minor principal axis.

$I_x = 999$; $r_x = 6.14$; $I_y = 362$; $r_y = 3.70$

Check width/thickness ratio, b/t, of the compression elements in the cross section(see LRFD pages 6-31,32):

1. For the flange(**unstiffened element**), local buckling does not occur before column buckling occurs provided:

$$\left[\frac{b}{t} = \frac{0.5b_f}{t_f} = 10.2\right] \le \left[\frac{95}{\sqrt{F_y}} = \frac{95}{\sqrt{36}} = 15.8\right]$$

Since 10.2≤15.8, local buckling of the flange does not occur before column buckling occurs.

2. For the web(**stiffened element**), local buckling does not occur before column buckling occurs provided:

$$\left[\frac{b}{t} = \frac{h_c}{t_w} = 25.9\right] \le \left[\frac{253}{\sqrt{F_y}} = \frac{253}{\sqrt{36}} = 42.2\right]$$

Since 25.9≤42.2, local buckling of the web does not occur before column buckling occurs.

The formulas in Figure 4.5 are valid since column buckling occurs first. Since KL = 20 feet for both principal axes and $r_y < r_x$, only λ_c for the Y axis is needed.

107

$$\lambda_{cy} = \frac{(KL)_y}{\pi r_y} \sqrt{\frac{F_y}{E}} = \frac{20*12}{3.70\pi} \sqrt{\frac{36}{29000}} = 0.727$$

Since $(\lambda_{cy} = 0.727) \le 1.5$, inelastic column buckling for the cross section rotating about the Y axis governs the axial compression design strength.

For the author's preference of formulas in Figure 4.5:

$$\lambda_c^2 = \left[\lambda_{cy}^2 = 0.529 \right]$$

$$\frac{E_t}{E} = \frac{\lambda_c^2(0.658^{\lambda_c^2})}{0.877} = \frac{0.529*(0.658^{0.529})}{0.877} = 0.4835$$

$$P_{cr} = 0.877*\frac{\pi^2 E_t I_y}{(KL)_y^2}$$

$$P_{cr} = 0.877*\frac{\pi^2*0.4835*29000*362}{(20*12)^2} = 762.8$$

$$\phi P_{cr} = 0.85*762.8 = 648 \text{ kips}$$

For the actual LRFD formulas in Figure 4.5:

$$\lambda_c^2 = \left[\lambda_{cy}^2 = 0.529 \right] \le 1.5$$

$F_{cr} = (0.658^{\lambda_c^2})*F_y = (0.658^{0.529})*36 = 28.85 \text{ ksi}$

$P_{cr} = A_g*F_{cr} = 26.5*28.85 = 764.5$

$\phi P_{cr} = 0.85*764.5 = 650 \text{ kips}$

EXAMPLE 4.2_____

An A36 steel W14X90 section is used as an axially loaded compression member with $(KL)_x = 20$ ft and $(KL)_y = 10$ ft. In EXAMPLE 4.1 we proved that local buckling does not occur before column buckling occurs. Therefore, use the formulas in Figure 4.5 to find the axial compression design strength.

SOLUTION -- A = 26.5; $I_x = 999$; $r_x = 6.14$; $I_y = 362$; $r_y = 3.70$

$$\lambda_{cx} = \frac{(KL)_x}{\pi r_x} \sqrt{\frac{F_y}{E}} = \frac{20*12}{6.14\pi} * \sqrt{\frac{36}{29000}} = 0.438$$

$$\lambda_{cy} = \frac{(KL)_y}{\pi r_y} \sqrt{\frac{F_y}{E}} = \frac{10*12}{3.70\pi} * \sqrt{\frac{36}{29000}} = 0.364$$

The cross section rotates about the X axis when column buckling occurs since $\lambda_{cx} > \lambda_{cy}$; $\lambda_c = 0.438$ and $\lambda_c^2 = 0.192$

For the author's preference of formulas in Figure 4.5:

$$\frac{E_t}{E} = \frac{\lambda_c^2 (0.658^{\lambda_c^2})}{0.877} = \frac{0.192*(0.658^{0.192})}{0.877} = 0.202$$

$$P_{cr} = 0.877*\frac{\pi^2 EI}{(KL)^2} = 0.877*\frac{\pi^2*0.202*29000*999}{(20*12)^2} = 880$$

$$\phi P_{cr} = 0.85*880 = 748 \text{ kips}$$

For the actual LRFD formulas in Figure 4.5:

$$[\lambda_c^2 = 0.192] \leq 1.5$$

$$F_{cr} = (0.658^{\lambda_c^2})*F_y = (0.658^{0.192})*36 = 33.2 \text{ ksi}$$

$$P_{cr} = A_g*F_{cr} = 26.5*33.2 = 880$$
$$\phi P_{cr} = 0.85*880 = 748 \text{ kips}$$

EXAMPLE 4.3_____

Solve EX 4.1 using LRFD page 2-20.

SOLUTION--See LRFD page 2-3 and read it. On LRFD page 2-15 add the following two sentences:

1. All sections listed satisfy LRFD Table B5.1 for axial compression with the exception of W14X43 for $F_y = 50$ ksi.

2. The single dagger and double dagger footnotes pertain to the b/t ratio for beam action if a section is used as a beam-column.

On LRFD page 2-20, enter at $(KL)_y = 20$ ft; find, $\phi P_{cr} = 650$ kips.

In EXAMPLE 4.1 we found $\phi P_{cr} = 650$ kips which agrees with the value in the table on LRFD page 2-20. Therefore, EXAMPLE 4.1 shows how a value in the table on LRFD page 2-20 was obtained.

EXAMPLE 4.4_____

Solve EX 4.2 using LRFD page 2-20.

SOLUTION--Enter at $(KL)_y = 10$ ft; find, $\phi P_{cry} = 767$ kips.
Since the table does not give column design strength values for the X axis, we must show that the following approach can be used to find the column design strength for the X axis. At the bottom of the page, find $r_x/r_y = 1.66$ for W14X90. Enter the table at $\dfrac{(KL)_x}{\frac{r_x}{r_y}} = \dfrac{20}{1.66} = 12.05$ ft;

use linear interpolation to find:

$\phi P_{crx} = 749 - 0.05*(749-738) = 748.45$ kips

Then, $\phi P_{cr} = 748$ kips (smaller of ϕP_{crx} and ϕP_{cry})

In EXAMPLE 4.2 we found $\phi P_{cr} = 748$ kips which agrees with the value obtained from the table on LRFD page 2-20. Therefore, we have shown that the described approach for using the column table to find the column design strength for the X axis is correct.

EXAMPLE 4.5_____

The required design strength is $P_u = 300$ kips.
Find the lightest:
    a) W14
    b) W12
    c) W10
    d) W8
of A36 steel that satisfies the LRFD Specs. for axial compression when $(KL)_x = 20$ ft and $(KL)_y = 10$ ft.

SOLUTION--The design requirement is: $\phi P_{cr} \geq (P_u = 300 \text{ kips})$

See LRFD page 2-15. Review the note that the author had the reader to add on this page(the note was given in EXAMPLE 4.3). Since A36 steel is being used and all sections satisfy LRFD Table B5.1, we start by selecting a section that satisfies the design requirement for the Y axis. For this selected section we find the $\frac{r_x}{r_y}$ ratio listed at the bottom of the column table. If $\frac{(KL)_x}{\frac{r_x}{r_y}} > (KL)_y$, column buckling occurs with the section rotating about the X axis and we must enter the column table with an assumed value for $\frac{(KL)_x}{\frac{r_x}{r_y}}$ in order to choose a section that satisfies the design requirement. Assume $\frac{r_x}{r_y}$ will be the same as it was for the section selected for the Y axis. If $\frac{r_x}{r_y}$ for the selected section differs significantly from the assumed value, compute a revised value of $\frac{(KL)_x}{\frac{r_x}{r_y}}$ to use in making the next selection.

a) select the lightest W14 that satisfies the design requirement
   On LRFD page 2-21, for $F_y = 36$ ksi, enter at $(KL)_y = 10$ ft and find the smallest value that exceeds 300 kips. If the Y axis governs the column design strength, a W14X43 ($\phi P_{cry} = 312$) ≥ 300 is the lightest choice. For a W14X43, $\frac{r_x}{r_y} = 3.08$ is found at the bottom of the table and is used to compute the following assumed value for entering the table to make the selection that satisfies the design requirement for the X axis:
   $$\frac{(KL)_x}{\frac{r_x}{r_y}} = \frac{20}{3.08} = 6.49 \text{ ft}$$
   W14X43 ($\phi P_{crx} = 350$) > ($\phi P_{cry} = 312$) ≥ 300 is the lightest W14 that satisfies the design requirement. It should be noted that the

assumed value of $r_x/r_y = 3.08$ used in entering the table at 6.49 ft was the correct value for the section selected for the X axis.

b) select the lightest W12 that satisfies the design requirement
   On LRFD page 2-25, for $F_y = 36$ ksi:

   enter at $(KL)_y = 10$ ft and find: W12X45 $(\phi P_{cry} = 330) \geq 300$

   enter at $\dfrac{(KL)_x}{\dfrac{r_x}{r_y}} = \dfrac{20}{2.65} = 7.55$ ft and find:

   W12X45 $(\phi P_{crx} = 360) > (\phi P_{cry} = 330) \geq 300$

c) select the lightest W10 that satisfies the design requirement
   On LRFD page 2-27, for $F_y = 36$ ksi:

   enter at $(KL)_y = 10$ ft and find: W10X45 $(\phi P_{cry} = 337) \geq 300$

   Enter at $\dfrac{(KL)_x}{\dfrac{r_x}{r_y}} = \dfrac{20}{2.15} = 9.30$ ft and find:

   W10X45 $(\phi P_{crx} = 346) > (\phi P_{cry} = 337) \geq 300$

d) select the lightest W8 that satisfies the design requirement
   On LRFD page 2-28, for $F_y = 36$ ksi:

   enter at $(KL)_y = 10$ ft and find: W8X48 $(\phi P_{cry} = 362) \geq 300$

   enter at $\dfrac{(KL)_x}{\dfrac{r_x}{r_y}} = \dfrac{20}{1.74} = 11.5$ ft and find:

   W8X48 $(\phi P_{crx} = 342) \geq 300$

   NOTE: For W8X40 $(\phi P_{cry} = 298) \approx 300$,

   but $(\phi P_{crx} = 280) < 300$
   W8X40 does not satisfy the design requirement.
   Lightest W8 choice is:

   W8X48 $(\phi P_{crx} = 362) > (\phi P_{cry} = 342) \geq 300$

EXAMPLE 4.6 _____

For the plane frame in Figure 4.13a, the reader should be able to make a rough sketch of the deflected shape due to the wind loads shown assuming that $W_2 > W_1$. For discussion purposes, assume that the member sizes in Figure 4.13a are those shown in Figure 4.13b and that due to the wind loads the members bend about their X axis in the deflected shape.

Assumptions 1 to 3 and 5 to 8 on LRFD page 6-152 were made in obtaining the nomograph on LRFD page 6-153. See Figure C-C2.1 on LRFD page 6-151. For an unbraced plane frame, the in-plane deflected shape due to the buckling loads which are applied joint loads in the gravity direction is assumed(see assumption 5) to have the same joint rotation directions and the same points of inflection as the deflected shape of the same structure due to wind loads applied at the joints and applied parallel to the wind direction.

Use the nomograph on LRFD page 6-153 and the correct definition of G (see text Eqns(4.16, 4.15, 4.14)) to find $(KL)_x$ for members 1 to 5 in the unbraced plane frame shown in Figure 4.13.

Also find ϕP_{crx} for members 1 to 5.

SOLUTION--W12X120 $I_x = 1070$; $r_x = 5.51$
W30X173 $I_x = 8200$; W30X116 $I_x = 4930$

Since $\frac{E_t}{E}$ is unknown in the definition of G(see Eqn(4.16)), **we must use an iterative procedure** in which we assume a value of $\frac{E_t}{E}$ and check the assumed value as illustrated in the Member 1 calculations shown below.

If $\lambda_c \leq 1.5$, inelastic column buckling occurs. Since all columns in this example are W12X120, for numerical convenience Eqn(4.14) and the design strength definition can be written as:

$$\lambda_{cx} = \left[\frac{L_x}{\pi r_x} \sqrt{\frac{F_y}{E}} \right] * K_x$$

$$\phi P_{crx} = 0.85 * \frac{E_t}{E} * \left[0.877 * \frac{\pi^2 E I_x}{L_x^2} \right] * \frac{1}{K_x^2}$$

113

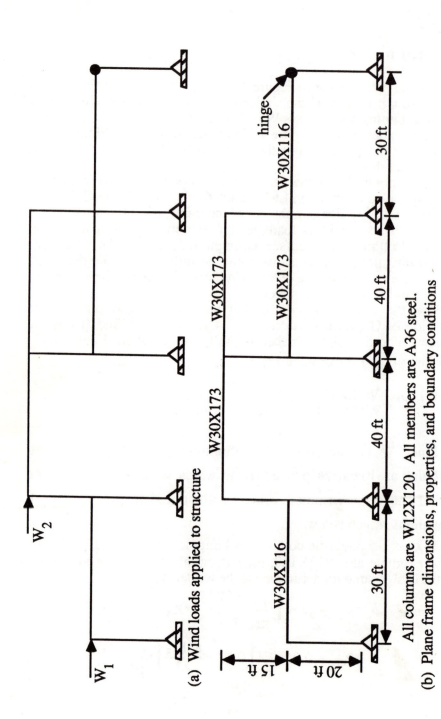

(a) Wind loads applied to structure

All columns are W12X120. All members are A36 steel.

(b) Plane frame dimensions, properties, and boundary conditions

Figure 4.13 An unbraced plane frame

114

For L = 15 ft = 180 in. (Members 3 and 5):

$$\lambda_{cx} = 0.366373*K_x; \quad \phi P_{crx} = \frac{7046*\frac{E_t}{E}}{K_x^2}$$

For L = 20 ft = 240 in. (Members 1,2,4):

$$\lambda_{cx} = 0.488497*K_x; \quad \phi P_{crx} = \frac{3963*\frac{E_t}{E}}{K_x^2}$$

It should be emphasized that the common factors shown above for each member were obtained using the actual precision of the author's electronic calculator. These common factors were recorded in the text with the appropriate precisions needed in the following calculations.

Member 1

G_{bottom}= 10 (use value recommended on LRFD page 6-153 for a hinged support).

At the top end, using Eqn(4.16): $G_{top} = \dfrac{\frac{E_t}{E}*\frac{1070}{20}}{\frac{1*4930}{30}} = 0.326*\frac{E_t}{E}$

First iteration cycle:

Assume elastic buckling occurs; that is, assume $\frac{E_t}{E} = 1$.

On LRFD page 6-153, for G_{top} = 0.326 and G_{bottom} = 10 find $K_x = 1.75$ (for elastic buckling).

Check our assumption which was that $\frac{E_t}{E} = 1$:

$[\lambda_{cx} = 0.488497*1.75 = 0.855] \leq 1.5;$ *inelastic buckling occurs.*

$\lambda_{cx}^2 = 0.7308; \left[\frac{E_t}{E} = \frac{0.7308*(0.658^{0.7308})}{0.877} = 0.614\right] \neq 1$

as we assumed. If we wish to find the correct value of ϕP_{crx}, we must continue the iteration procedure until the computed $\frac{E_t}{E}$ = assumed $\frac{E_t}{E}$. However, the author will show later that: at the end of the first iteration cycle, if we use the computed $\frac{E_t}{E}$ value and K_x found for $\frac{E_t}{E}$ = 1 in the above definition of ϕP_{crx} to obtain $\phi P_{crx} = \dfrac{3963*0.614}{(1.75)^2} = 795$ kips which is an approximate, lower bound value and is conservative to use for design purposes.

Alternatively, we can use LRFD page 2-23 to find this ϕP_{crx} :

enter at $\dfrac{(KL)_x}{\dfrac{r_x}{r_y}}$ $= \dfrac{1.75*20}{1.76} = 19.9$ ft and by interpolation find:

$\phi P_{crx} = 793 + 0.1*24 = 795$ kips

Second iteration cycle:
    In the first iteration cycle, we assumed $E_t/E = 1$ and found $E_t/E = 0.614$ based on K_x obtained for $E_t/E = 1$. In the second iteration cycle, we know that $E_t/E < 0.614$ and the author chooses to assume $E_t/E = 0.6$ for illustration purposes.

$G_{bottom} = 10$; $G_{top} = 0.326*(E_t/E) = 0.326*0.6 = 0.196$; $K_x = 1.72$
$\lambda_{cx} = 0.488497*1.72 = 0.840$; $\lambda^2_{cx} = 0.706$

$[E_t/E = 0.706*(0.658^{0.706})/0.877 = 0.599] \approx [0.6$ assumed$]$
$\phi P_{crx} = 3963*0.599/(1.72)^2 = 802$ kips
This is the correct design strength since the assumed E_t/E value was correct.

Summary for Member 1:
First iteration cycle, assumed $E_t/E = 1$
                     computed $E_t/E = 0.614$; $\phi P_{crx} = 795$ kips
Converged iteration cycle, assumed $E_t/E = 0.6$
                     computed $E_t/E = 0.599$; $\phi P_{crx} = 802$ kips
    Therefore, we have shown that the design strength value obtained in the first iteration cycle and based on K_x for $E_t/E = 1$ is an approximate, lower bound estimate of the correct design strength. Note that $802/795 = 1.009$ and the correct design strength for this member is only 0.9% larger than the lower bound estimate of the design strength.
    It must be emphasized that for any other column in this example or for any column in another frame, we only know that the correct design strength is larger than the lower bound estimate found in the first iteration cycle by using the elastic effective length. That is, in general we do **not** know what the correct factor is to multiply the lower bound estimate by to obtain the correct design strength.

Member 2

$G_{bottom}= 10$

$G_{top}= [(E_t/E)_2*(1070/20) + (E_t/E)_3*(1070/15)]/[1*(4930/30)]$

First iteration cycle:

Assume $E_t/E = 1$ for members 2 and 3. $G_{top} = 0.760$; $K_x = 1.85$

$\lambda_{cx} = 0.488497*1.85 = 0.904$; $\lambda_{cx}^2 = 0.817$

$(E_t/E) = 0.817*(0.658^{0.817})/0.877 = 0.662$

Lower bound estimate of $\phi P_{crx} = 3963*0.662/(1.85)^2 = 767$ kips

Note that $(E_t/E)_3$ is involved in the definition of G and we will not be able to obtain a reliable estimate of $(E_t/E)_3$ until we perform the first iteration cycle for Member 3. Therefore, at any joint where there are two columns i and j, if both columns buckle inelastically, the iteration cycles for Members i and j are not independent. However, we can independently find the lower bound estimate of the design strength for each column using the elastic effective length of each column. Fortunately, we can avoid additional iterations since we can accept the lower bound estimate for each column as the column design strength.

Member 3

First iteration cycle:

elastic $G_{bottom}= 0.760$ (same as G_{top} for Member 2)

elastic $G_{top}= (1070/15)/[1*(8200/40)] = 0.348$

elastic $K_x = 1.17$

$\lambda_{cx} = 0.366373*1.17 = 0.429$; $\lambda_{cx}^2 = 0.184$; $(E_t/E) = 0.194$

Lower bound estimate of $\phi P_{crx} = 7046*0.194/(1.17)^2 = 999$ kips

Member 4

$G_{bottom}= 1$ (use value recommended on LRFD page 6-153 for a fixed support).

First iteration cycle:

elastic $G_{top} = [1070/20 + 1070/15]/[1*(8200/40) + 0.5*(4930/30)]$
            $= 0.435$

elastic $K_x = 1.22$

$\lambda_{cx} = 0.488497*1.22 = 0.596$; $\lambda_{cx}^2 = 0.355$; $E_t/E = 0.349$

Lower bound estimate of $\phi P_{crx} = 3963*0.349/(1.22)^2 = 929$ kips

Member 5

First iteration cycle:

elastic $G_{bottom} = 0.435$ (same as G_{top} for Member 4)
elastic $G_{top} = (1070/15)/[1*(8200/40)] = 0.348$
elastic $K_x = 1.13$

$\lambda_{cx} = 0.366373*1.13 = 0.414; \quad \lambda^2_{cx} = 0.171; \quad E_t/E = 0.182$

Lower bound estimate of $\phi P_{crx} = 7046*0.182/(1.13)^2 = 1004$ kips

EXAMPLE 4.7 _____

Correctly solve EXAMPLE 3 on LRFD page 2-6 using the:
    1) author's approach
    2) approach recommended on LRFD page 2-6

SOLUTION using the author's approach
    The lightest acceptable W12 for $(KL)_y = 15$ ft. is found from the column table on LRFD page 2-23: W12X120 ($\phi P_{cry} = 908$) ≥ 840. Try W12X120 for each column in the figure on LRFD page 2-6.

First iteration cycle:
    $G_{bottom} = 10$
elastic $G_{top} = (1070/15)/[1*(375/20)] = 3.80$
elastic $K_x = 2.42$

$\lambda_{cx} = [2.42*15*12)/(5.51\pi)]*\sqrt{36/29000} = 0.8866; \quad \lambda^2_{cx} = 0.7861$

$E_t/E = 0.7861*(0.658^{0.7861})/0.877 = 0.645$
Lower bound estimate:

$\phi P_{crx} = 0.85*0.645*0.877*\pi^2*29000*1070/(2.42*15*12)^2 = 776$ kips

Second iteration cycle:
Assume $E_t/E = 0.53$
$G_{top} = 3.80*(E_t/E) = 3.80*0.53 = 2.01; \quad K_x = 2.10$
$\lambda_{cx} = 0.7694; \quad \lambda^2_{cx} = 0.5920; \quad [(E_t/E) = 0.527] \approx [0.53 \text{ assumed}]$

$\phi P_{crx} = 0.85*0.527*0.877*\pi^2*29000*1070/(2.10*15*12)^2 = 842$ kips
$[\phi P_{cr} = 842] \geq [P_u = 840];$ therefore, W12X120 is okay to use.

Note that $842/776 = 1.085$; that is, in this example the correct value of the design strength is 8.5% larger than the lower bound estimate found in the first iteration cycle. The reason that the lower bound estimate is not closer to the correct design strength is a W16X31 girder is not a realistic choice in an actual design situation. See the author's EXAMPLE 4.8 for a more realistic choice of girder for the figure on LRFD page 2-6.

SOLUTION using recommended approach on LRFD page 2-6

STEP 2 $\dfrac{P_u/\phi}{A} = \dfrac{840/0.85}{35.3} = 28.0$ ksi

STEP 3 Enter LRFD TABLE A page 2-9 for $F_y = 36$ ksi

at $\dfrac{P_u/\phi}{A}$ =28.0 ksi

find an estimate of $E_t/E = 0.532$
NOTE: Heading for column 1 of TABLE A should be

$$\dfrac{P_u/\phi}{A}$$

STEP 4 is okay as shown on LRFD page 2-6.
STEP 5 $G_{top} = 3.80*0.532 = 2.02$
STEP 6 $K_x = 2.10$

STEP 7 The author chooses to use LRFD page 2-23 to find ϕP_{crx}
On LRFD page 2-23 for W12X120 and for $F_y = 36$ ksi,

enter at $\dfrac{(KL)_x}{\dfrac{r_x}{r_y}} = 2.10*15/1.76 = 17.9$ ft

and find [$\phi P_{crx} = 841 + 0.1*23 = 843$ kips] $\geq$ [$P_u = 840$] and 843 kips is the same as the 842 kips found by the author's approach. W12X120 is okay to use and is the lightest acceptable W12 section.

EXAMPLE 4.8 _____

In the figure on LRFD page 2-6, change the girder to a W24X76 for which $I_x = 2100$. For the W12X120 $F_y = 36$ ksi columns, find the column design strength. Compare to the EXAMPLE 4.7 solution.

SOLUTION using the author's approach
$G_{bottom} = 10$

$G_{top} = [(E_t/E)*1070/15]/[1*(2100/20)] = 0.6794*(E_t/E)$

Iteration cycle 1:
Assume $E_t/E = 1$; elastic $K_x = 1.85$

$\lambda_{cx} = 0.366373*1.85 = 0.678;\quad \lambda^2_{cx} = 0.4594$

$E_t/E = 0.4594*(0.658^{0.4594})/0.877 = 0.432$
Lower bound estimate:

$\phi P_{crx} = 0.85*0.432*0.877*\pi^2*29000*1070/(1.85*15*12)^2 = 889$ kips

Iteration cycle 2:
Assume $E_t/E = 0.4;\quad G_{top} = 0.6794*0.4 = 0.272;\quad K_x = 1.75$

$\lambda_{cx} = 0.641;\quad \lambda^2_{cx} = 0.411;\quad [E_t/E = 0.395] \approx (0.4$ assumed$)$

$\phi P_{crx} = 909$ kips
Note that $909/889 = 1.0225$ and the correct column design strength is only 2.2% larger than the lower bound estimate.

SOLUTION using approach recommended on LRFD page 2-6
Enter TABLE A on LRFD page 2-9 for $F_y = 36$ ksi

at $\dfrac{P_u/\phi}{A} = \dfrac{840/0.85}{35.3} = 28.0$ ksi and find an estimate of $E_t/E = 0.532$

$G_{top} = 0.6794*0.532 = 0.361;\quad K_x = 1.77$

$\dfrac{(KL)_x}{\dfrac{r_x}{r_y}} = \dfrac{1.77*15}{1.76} = 15.1$ ft

On LRFD page 2-23, $\phi P_{crx} = 908 - 0.1*12 = 907$ kips

    Comparison of solutions for EXAMPLE 4.7 and EXAMPLE 4.8 reveals that by using a more realistic girder size of W24X76 instead of a W16X31, the column design strength for the frame on LRFD page 2-6 is increased by a factor of $909/842 = 1.0796$ which is an increase of about 8.0%. Furthermore, for a more realistic girder size, the lower bound estimate of the design strength obtained by using the elastic $(KL)_x$ value is very nearly equal to the correct design strength.

EXAMPLE 4.9_____

In the figure on LRFD page 2-6, change the girder to a W12X22 for which $I_x = 156$. For the W12X120 $F_y = 36$ ksi columns, find the column design strength. Compare the solution to the EXAMPLE 4.7 solution.

SOLUTION using the author's approach
$G_{bottom} = 10$
$G_{top} = [(E_t/E)*1070/15]/[1*(156/20)] = 9.15*(E_t/E)$

Iteration cycle 1:
Assume $E_t/E = 1$; elastic $K_x = 2.95$
$\lambda_{cx} = 0.366373*2.95 = 1.081$; $\lambda^2_{cx} = 1.168$

$E_t/E = 1.168*(0.658^{1.168})/0.877 = 0.817$
Lower bound estimate:
$\phi P_{crx} = 0.85*0.817*0.877*\pi^2*29000*1070/(2.95*15*12)^2 = 661$ kips

Iteration cycle 2:
Assume $E_t/E = 0.77$; $G_{top} = 9.15*0.77 = 7.05$; $K_x = 2.78$
$\lambda_{cx} = 1.019$; $\lambda^2_{cx} = 1.037$; $[E_t/E = 0.766] \approx (0.77$ assumed$)$

$\phi P_{crx} = 698$ kips
Note that $698/661 = 1.056$ and the correct column design strength is 5.6% larger than the lower bound estimate.

SOLUTION using approach recommended on LRFD page 2-6
Enter TABLE A on LRFD page 2-9 for $F_y = 36$ ksi

at $(P_u/\phi)/A = (840/0.85)/35.3 = 28.0$ ksi
and find an estimate of $E_t/E = 0.532$
$G_{top} = 9.15*0.532 = 4.87$; $K_x = 2.57$

$\dfrac{(KL)_x}{\dfrac{r_x}{r_y}} = 2.57*15/1.76 = 21.9$ ft

On LRFD page 2-23, $\phi P_{crx} = 743 + 0.1*50/2 = 746$ kips

Note that the correct value of $\phi P_{crx} = 628$ kips found in the author's approach).and $746/628 = 1.19$. For this example,the LRFD recommended approach **is not conservative** and over estimates the design strength by 19%.

If we enter TABLE A on LRFD page 2-9 at $P_u/A = 840/35.3 = 23.8$ ksi and use the estimate of $E_t/E = 0.745$

$G_{top} = 9.15*0.745 = 6.82;$ $K_x = 2.75$

$$\frac{(KL)_x}{\dfrac{r_x}{r_y}} = 2.75*15/1.76 = 23.4 \text{ ft}$$

On LRFD page 2-23, $\phi P_{crx} = 692 + 0.6*51/2 = 707$ kips

Note that $707/628 = 1.13$ and this approach <u>over estimates</u> the design strength by 13%.

This example shows that the LRFD recommended approach does not give a conservative estimate of the design strength when the estimate of design strength obtained by the LRFD recommended approach is less than P_u. However, <u>the author's approach **always** gives a lower bound estimate of the correct design strength.</u>

EXAMPLE 4.10 _____

A pair of A36 steel L4X3X3/8 with long legs back to back is used as a 6 ft. long compression member in a truss. The long leg of each angle is adequately welded at the member ends to a 3/8 in. thick gusset plate.

If the structural designer requires a spacer plate of 3/8 in. thickness to be inserted between the long legs of the angles at each third point of the member length and requires each spacer plate to be either welded or bolted with a fully tightened bolt to the long leg of each angle, the column design strength for this built up section member can be found from LRFD page 2-64. In a truss analysis, each member is assumed to be pinned ended. Therefore, using $(KL)_x = (KL)_y = 6$ ft and $F_y = 36$ ksi, on LRFD page 2-64 we find $\phi P_{cr} = 123$ kips (smaller of $\phi P_{crx} = 128$ and $\phi P_{cry} = 123$ kips) . See and read LRFD page 2-46.

This is the only example in which the author shows how to use the LRFD Specification equations to find the column design strength of a built up section member. Therefore, the purpose of this example is to illustrate the computations for each of the potential column design strength candidates assuming either welded or fully tightened bolted

connectors are provided at each of the following intermediate connector spacing intervals along the member:
1) a = 36 in. (1 connector)
2) a = 24 in. (2 connectors)
SOLUTION--See LRFD page 6-32:

Since (b/t = 4/0.375 = 10.7) ≤ (76/$\sqrt{36}$ = 12.7), local buckling does not govern the column design strength and LRFD Appendix B is not applicable. Therefore, for the built up section being investigated in this example, the column design strength is governed either by one of the flexural column buckling modes or by the flexural-torsional column buckling mode.

For the flexural column buckling modes:
1. For *behavior as two single angles*, LRFD E2(page 6-39) is applicable.
   See LRFD page 1-56,57 for the properties of a single angle:
   L4X3X3/8: A = 2.48; r_{min} = r_z = 0.644

   $I_z = A*r_z^2$ = 2.48*(0.644) = 1.03 is the moment of inertia for the

   minor principal axis of this single angle.
   See the figure on LRFD page 2-64. At each intermediate connector location, each long leg of each angle is connected either by welds or by a fully tightened bolt to a spacer plate which is not prevented from translating in each of the double angle principal axis directions. However, at each member end and at each intermediate connector, each angle is prevented from translating in each of the single angle principal axis directions, w and z, since w and z are not parallel to either of the principal axes, X and Y, of the double angle section. Therefore, for behavior as two single angles, flexural column buckling about the z axis of each angle can occur if the intermediate connector(s) are too far apart. At the member ends and at the intermediate connector(s), the amount of rotational resistance about the z axis of each single angle section is small and cannot be easily determined. Therefore, we conservatively assume (KL)$_z$ = a in the following calculations.
   a) for (KL)$_z$ = a = 36 in., the *flexural column buckling mode* for *each single angle* is two half sine waves and their length is 36 in.

   $$\left[\lambda_{cz} = \frac{36}{0.644\pi}*\sqrt{\frac{36}{29000}} = 0.6269\right] \le 1.5$$

   $$\lambda_{cz}^2 = 0.3930$$

Using the author's approach:

$E_t/E = 0.3930*(0.658^{0.393})/0.877 = 0.3802$

For one angle:

$\phi P_{crxz} = 0.85*0.3802*0.877*\pi^2*29000*1.03/(36)^2$
$= 64.5$ kips

For two single angles: $\phi P_{crz} = 2*64.5 = 129$ kips

Using the LRFD approach for two single angles:

$\phi P_{crz} = 2*[0.85*2.48*(0.658^{0.393})*36] = 129$ kips

b) for $(KL)_z = a = 24$ in., the *flexural column buckling mode* for *each single angle* is three half sinewaves and each half sine wave length is 24 in.

For two single angles, using the LRFD approach

$\lambda_{cz} = (24/36)0.6269 = 0.4180;\ \lambda^2_{cz} = 0.1747$

$\phi P_{crz} = 0.85*2*2.48*(0.658^{0.1747})*36 = 141$ kips

2. For *double angle section behavior*, see LRFD pages 1-90,91 for the following properties:

$A = 4.97;\ I_x = 7.93;\ r_x = 1.26;\ y = 1.28;\ r_y = 1.31$
$I_y = A*r_y^2 = 4.97*(1.31)^2 = 8.53$

a) for *bending about the X axis*, LRFD E2(page 6-39) is applicable and the *flexural column buckling mode* is a half sine wave whose length is 72 in.

$$\left[\lambda_{cx} = \frac{72}{1.26\pi}*\sqrt{\frac{36}{29000}} = 0.6409\right] \leq 1.5$$

$\lambda^2_{cx} = 0.4107$

$\phi P_{crx} = 0.85*4.97*(0.658^{0.4107})*36 = 128$ kips

b) for bending about the Y axis, LRFD E2(page 6-39) and LRFD E4 (page 6-40--item b, for this example) are applicable. The *flexural column buckling mode* is a half sine wave whose length is 72 in., but there is another phenomenon which the author must explain before the buckling load calculations can be made.

124

The flexural column buckling behavior of two individual b X d rectangular sections that are not interconnected in any way is shown in Figure 4.14a for bending about the y axis. Note that slippage between the two sections occurs everywhere along the member length except at midlength of the member. The maximum slippage occurs at each member end. If the exterior edges of the slippage plane are sealed by a weld, the two individual sections are fully interconnected along the member length and form a built up section. The flexural column buckling behavior of the built up section is as shown in Figure 4.14b since the welds prevent any slippage between the two joined sections. Therefore, the welds are subjected to shear when the member bends about the y axis. However, the welds are not subjected to shear when the member bends about the x axis.

For ease in vizualizing the following discussion for the double angle section bending about the y axis in the example problem, assume that each intermediate connector is a fully tightened bolt. The length direction of the bolt is parallel to the X axis in the figure on LRFD page 2-64. Since bending is about the Y axis in the figure on LRFD page 2-64, due to the internal bending moment which exists when flexural column buckling occurs, one angle is in tension and the other angle is in compression whereas both angles are in compression due to the axial load. Therefore, when Y axis flexural column buckling occurs, the member end connections and the intermediate connectors(except for the one at midlength of the member) are subjected to shear.

b) **continued** for *flexural column buckling mode*
   1) for a = 36 in.: $(a/r_z = 36/0.644 = 55.9) > 50$
      LRFD Eqn(E4-2) page 6-40 is applicable.

$$\left(\frac{KL}{r}\right)_m = \sqrt{\left(\frac{KL}{r}\right)_y^2 + \left(\frac{a}{r_z} - 50\right)^2}$$

$$\left(\frac{KL}{r}\right)_y = \frac{72}{1.31} = 54.96$$

$$\left(\frac{KL}{r}\right)_m = \sqrt{(54.96)^2 + (55.90 - 50)^2} = 55.3$$

$$\lambda_{cy} = \frac{55.3}{\pi}\sqrt{\frac{36}{29000}} = 0.6199; \quad \lambda_{cy}^2 = 0.3843$$

$$\phi P_{cry} = 0.85*4.97*(0.658^{0.3843})*36 = 129.5 \text{ kips}$$

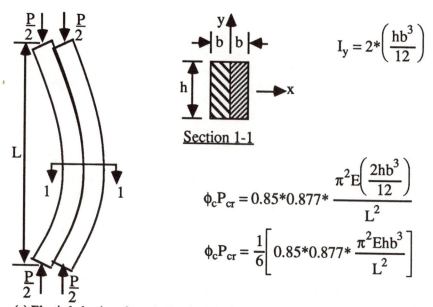

$$I_y = 2*\left(\frac{hb^3}{12}\right)$$

Section 1-1

$$\phi_c P_{cr} = 0.85*0.877*\frac{\pi^2 E\left(\frac{2hb^3}{12}\right)}{L^2}$$

$$\phi_c P_{cr} = \frac{1}{6}\left[0.85*0.877*\frac{\pi^2 Ehb^3}{L^2}\right]$$

(a) Elastic behavior of two individual b X d pieces(no interconnection)

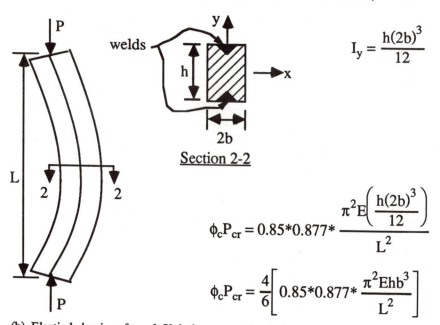

$$I_y = \frac{h(2b)^3}{12}$$

Section 2-2

$$\phi_c P_{cr} = 0.85*0.877*\frac{\pi^2 E\left(\frac{h(2b)^3}{12}\right)}{L^2}$$

$$\phi_c P_{cr} = \frac{4}{6}\left[0.85*0.877*\frac{\pi^2 Ehb^3}{L^2}\right]$$

(b) Elastic behavior of two b X d pieces completely interconnected

Figure 4.11 Flexural column buckling for y-axis bending

2) for a = 24 in.: $(a/r_z = 24/0.644 = 37.3) < 50$

$(KL/r)_m = (KL/r)_y = 72/1.31 = 54.96$

$$\lambda_{cy} = \frac{55.96}{\pi}\sqrt{\frac{36}{29000}} = 0.616; \quad \lambda^2_{cy} = 0.3800$$

$\phi P_{cry} = 0.85*4.97*(0.658^{0.3800})*36 = 129.7$ kips

b) **continued** -- For the *flexural-torsional column buckling mode* of the double angle section, LRFD Appendix E3(page 6-90) is applicable.

   $Q = 1$ since $(b/t = 4/0.375 = 10.7) \le (76/\sqrt{36} = 12.7$ from LRFD TABLE B5.1 page 6-32). The shear center is on the Y axis at midthickness of the shorter angle legs. Twisting of the cross section is prevented only at the member ends.

$x_0 = 0; \quad y_0 = y - \frac{t}{2} = 1.28 - \frac{0.375}{2} = 1.0925$

$x^2_0 = 0; \quad y^2_0 = 1.19$

$$\bar{r}^2_0 = x^2_0 + y^2_0 + \frac{I_x + I_y}{A}$$

$\bar{r}^2_0 = 0 + 1.19 + (7.93 + 8.53)/4.97 = 4.50$

$H = 1 - (x^2_0 + y^2_0)/\bar{r}^2_0 = 1 - (0 + 1.19)/4.50 = 0.736$

$\bar{r}_0 = \sqrt{4.50} = 2.12$

   On LRFD page 1-160 for a pair of L4X3X3/8 with long legs back to back with a 3/8 in. separation, we find: $\bar{r}_0 = 2.12$ which agrees with 2.12 calculated above by the author and $H = 0.735$ which very nearly agrees with 0.736 by calculated above by the author.

1. for a = 36 in. (*flexural-torsional column buckling mode.*)

$$F_{ey} = \frac{\pi^2 E}{(KL/r)^2_m} = \pi^2 * 29000/(55.3)^2 = 93.59 \text{ ksi}$$

$$F_{ez} = \left(\frac{\pi^2 E C_w}{(KL)^2_z} + GJ\right)\frac{1}{A\bar{r}^2_0}$$

On LRFD page 6-16, we find: G = 11,200 ksi.
See LRFD page 1-145 for single L4X3X3/8 torsional properties and double them for a pair of angles that are not torsionally interconnected except at the member ends:
$C_w = 2*0.114 = 0.228$; $J = 2*0.123 = 0.246$

$$F_{ez} = [\pi^2 * 29000 * 0.228/(72)^2$$
$$+ 11200*0.246]/[4.97*4.50]$$

$F_{ez} = 123.8$ ksi
From LRFD page 6-91 item b:

$$F_e = \frac{F_{ey} + F_{ez}}{2H} * \left[1 - \sqrt{1 - \frac{4F_{ey}F_{ez}H}{(F_{ey}+F_{ez})^2}}\right]$$

$$F_e = \frac{93.59+123.8}{2*0.736} * \left[1 - \sqrt{1 - \frac{4*93.59*123.8*0.736}{(93.59 + 123.8)^2}}\right]$$

$F_e = 69.8$ ksi

$$\lambda_e = \sqrt{F_y/F_e} = \sqrt{36/69.8} = 0.7182; \quad \lambda^2_e = 0.5159$$

$$\phi P_{cr} = 0.85*4.97*(0.658^{0.5159})*36 = 122.5 \text{ kips}$$

2. for a = 24 in. (*flexural-torsional column buckling mode.*)

$$F_{ey} = \pi^2 * 29000/(54.96)^2 = 94.76 \text{ ksi}$$

$$F_e = \frac{94.76+123.8}{2*0.736} * \left[1 - \sqrt{1 - \frac{4*94.76*123.8*0.736}{(94.76 + 123.8)^2}}\right]$$

$F_e = 70.3$ ksi

$$\lambda_e = \sqrt{\frac{36}{70.3}} = 0.7154; \quad \lambda^2_e = 0.5118$$

$$\phi P_{cr} = 0.85*4.97*(0.658^{0.5118})*36 = 122.8 \text{ kips}$$

Summary of author's computations and comparison to values given on LRFD page 2-64:

1. For *flexural column buckling behavior* as **two single angles** when bending is such that <u>connectors are **not** subjected to shear</u>:

   $[\phi P_{crx} = 128] = [128$ on LRFD page 2-64 at $(KL) = 6$ ft$]$

   This value is not a function of the intermediate connector spacing.

2. For *flexural column buckling behavior* as **two single angles** when bending is such that <u>connectors are subjected to shear</u>:

   a) for a = 36 in. (one intermediate connector)

   $[\phi P_{crz} = 129 \text{ kips}] > [\phi P_{crx} = 128] > [\phi P_{cry} = 122.5]$

   b) for a = 24 in. (two intermediate connectors)

   $[\phi P_{crz} = 141 \text{ kips}] > [\phi P_{crx} = 128] > [\phi P_{cry} = 122.8]$

3. For **double angle section behavior**, *flexural-torsional buckling mode* governs

   a) for a = 36 in. (one intermediate connector)

   $\phi P_{cry} = 122.5$ kips and LRFDpage 2-64 does not give a solution for one intermediate connector

   b) for a = 24 in. (two intermediate connectors)

   $[\phi P_{cry} = 122.8] = 123$ on LRFD page 2-64 at $(KL) = 6$ ft$]$

Therefore, the entries on LRFD page 2-64 were correctly computed for two intermediate connectors and the governing design strength is 123 kips which is due to the flexural-torsional buckling mode.

It should be noted that if the objective of this example had been to use the LRFD formulas to compute only the governing design strength **for two intermediate connectors**(a = 24 in.), the author would have only needed to do the following:

1. For the *flexural column buckling mode*, find ϕP_{cr} for the largest of the three slenderness ratios:

   a) for double angle behavior:
   $(KL/r)_x = 72/1.26 = 57.14$
   [LRFD page 6-39 is applicable]

   b) for built-up section behavior:
   $(KL/r)_m = (KL/r)_y = 72/1.31 = 54.96$
   [LRFD page 6-40 is applicable]

   c) for pair of single angles behavior:
   $(KL/r)_z = a/r_z = 24/0.644 = 37.3$
   [LRFD page 6-39 is applicable]

$(KL/r)_x$ governs and $\phi P_{crx} = 128$ kips

2. For the *flexural-torsional buckling mode* :
   a) for the singly symmetric double angle section,
      find $\phi P_{cr} = 123$ kips using LRFD pages 6-90,91(item b)
   b) for a pair of single angles, LRFD pages 6-90,91(item c) are applicable. NOTE: This solution was **not** obtained in the above example since the author knew from experience that
      ϕP_{cr} for this solution differs very little from ϕP_{crz} for text item 1c above[see LRFD page 6-201].

Therefore, the design strength is 123 kips(the smaller value obtained for flexural column buckling and for flexural-torsional column buckling). Flexural-torsional buckling governs for this example problem.

EXAMPLE 4.11_____

The top chord members of the truss in Figure 3.2 on page 3-5 are to be selected. The same pair of angles with long legs back to back and welded to 3/8 in. thick gusset plates is to be used for members 15 to 24. Select the lightest acceptable pair of A36 steel angles. Specify the number of intermediate connectors that are needed.

SOLUTION--In text Appendix A for members 15 to 24, the maximum axial compression force is found to be 122.7 kips in members 18 and 21 for Loading 7. Assume the bending moment given in the computer solution is negligible.

The length of the members is:

$$\left[(12 \text{ in./ft})*\sqrt{(30)^2+(2.5)^2} \right]/5 = 72.25 \text{ in.} = 6.02 \text{ ft.}$$

At $(KL)_x = (KL)_y = 6.02$ ft. on LRFD pages 2-67,66,65, we find that none of the listed sections is acceptable and on LRFD page 2-64 we find:

1. L4X3X3/8 ($\phi P_{cr} = 123$) $\geq$ ($P_u = 122.7$) and 17 lb/ft.
2. Two intermediate welded or fully tightened bolted connectors are needed. Therefore, the intermediate connector spacing must not exceed (72.25 in.)/(2 + 1) = 24.08 in.

PROBLEMS

4.1 For the column shown in Figure 4.2 on text page 4-2.1, find ϕP_{cr} for $F_y = 36$ ksi.

4.2 Find ϕP_{cr} for: W14X90 $F_y = 65$ ksi; $(KL)_x = (KL)_y = 20$ ft

4.3 Find ϕP_{cr} for:
W14X90 $F_y = 65$ ksi; $(KL)_x = 20$ ft; $(KL)_y = 10$ ft

4.4 Use LRFD page 2-20 and find find ϕP_{cr} for:
W14X90 $F_y = 50$ ksi; $(KL)_x = (KL)_y = 20$ ft

4.5 Use LRFD page 2-20 and find ϕP_{cr} for:
W14X90 $F_y = 50$ ksi; $(KL)_x = 20$ ft; $(KL)_y = 10$ ft

4.6 Given: $P_u = 550$ kips; $(KL)_x = 20$ ft; $(KL)_y = 10$ ft; $F_y = 50$ ksi
Find the lightest acceptable:
   a) W14
   b) W12
   c) W10
   d) W8

131

4.7 $F_y = 36$ ksi; All columns are W12X65. $(KL)_y = L$ for all columns. Girders are as shown in Figure P4.7. All members bend about the x-axis for in-plane frame buckling. For $E_t/E = 1$ in Eqn(4.16) on text page 4-7, use LRFD page 6-153 to find K_x.

Find the design compressive strength, ϕP_{cr}, for members 1 to 4 in Figure P4.7.

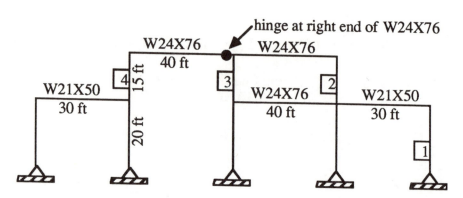

All columns are W12X65. All members are A36 steel.

Figure P4.7 An unbraced plane frame

4.8 A pair of L6X4X1/2 with long legs back to back and 3/4 in. thick separators spaced at intervals of 6 ft. 8 in. along the member length is used as a compression member in a truss.

$(KL)_x = (KL)_y = 20$ ft. Find ϕP_{cr} for $F_y = 36$ ksi.

4.9 A pair of L6X4X1/2 with long legs back to back and 3/4 in. thick separators spaced at intervals of 6 ft. 8 in. along the member length is used as a compression member in a truss.

$(KL)_x = (KL)_y = 20$ ft. Find ϕP_{cr} for $F_y = 65$ ksi.

4.10 A pair of C12X30 with tie plates spaced at intervals of 4 ft along the member length is used as a compression member in a truss. See Figure P4.10 for the cross section dimensions.

$(KL)_x = (KL)_y = 20$ ft. Find ϕP_{cr} for $F_y = 36$ ksi.

tie plates only exist at 4 ft. intervals
along the member length

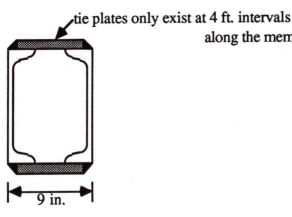

9 in.

Figure P4.10

4.11 $P_u = 360$ kips; $F_y = 36$ ksi; $(KL)_x = (KL)_y = 20$ ft.
Find the lightest pair of angles with 3/8 in. thick separators to serve as a compression member in a truss. Specify the minimum acceptable number of intermediate connectors and the maximum connector spacing.

4.12 Solve problem 4.11 for $F_y = 50$ ksi.

4.13 $P_u = 240$ kips; $F_y = 36$ ksi; $(KL)_x = (KL)_y = 16$ ft.
Find the lightest pair of angles with 3/8 in. thick separators to serve as a compression member in a truss. Specify the minimum acceptable number of intermediate connectors and the maximum connector spacing.

4.14 Solve problem 4.13 for $F_y = 50$ ksi.

4.15 $P_u = 240$ kips; $F_y = 36$ ksi; $(KL)_x = 8$ ft; $(KL)_y = 16$ ft.
Find the lightest pair of angles with 3/8 in. thick separators to serve as a compression member in a truss. Specify the minimum acceptable number of intermediate connectors and the maximum connector spacing.

133

4.16 Solve problem 4.15 for $F_y = 50$ ksi.

4.17 $P_u = 200$ kips; $F_y = 36$ ksi; $(KL)_x = (KL)_y = 10$ ft.
Find the lightest acceptable WT section to serve as a compression member in a truss.

4.18 Solve problem 4.17 for $F_y = 50$ ksi.

4.19 $P_u = 200$ kips; $F_y = 36$ ksi; $(KL)_x = 5$ ft; $(KL)_y = 10$ ft.
Find the lightest acceptable WT section to serve as a compression member in a truss.

4.20 Solve problem 4.19 for $F_y = 50$ ksi.

4.21 $P_u = 130$ kips; $F_y = 36$ ksi; $(KL)_x = 5$ ft; $(KL)_y = 10$ ft.
Find the lightest acceptable WT section to serve as a compression member in a truss.

4.22 Solve problem 4.21 for $F_y = 50$ ksi.

Chapter 5

BEAMS

5.1 INTRODUCTION

A *beam* is defined as any structural member that bends or/and twists due to the applied loads which do not cause an internal axial force to occur in the member. Therefore, an applied load on a beam cannot have any component parallel to the member length. Concentrated and distributed loads between the member ends, and member end moments are examples of applied loads on a beam.

For notational convenience on structural drawings and in structural design calculations, beams are also referred to as: *girders* (beams that support the most load in a floor system, for example; girders are the beams spaced at the largest interval in a floor system; the primary loads on girders are the reactions of other beams and/or possibly some columns); *floor beams* (beams that support joists); *joists* (the most closely spaced beams in a floor system; joists support the floor deck[concrete floor slab, for example]; steel joists may be either lightweight rolled sections or open-web joists[small trusses]); *roof beams* (beams that support purlins); *purlins* (most closely spaced beams in a roof system; purlins support the roof surface material and may be open-web joists, hot-rolled sections, or cold-formed sections); *spandrel beams* (beams supporting the outside edges of a floor deck and the outside walls of a building up to next floor level); *lintels* (beams spanning over window and door openings in a wall; a lintel supports the wall portion above a window or door opening); *girts* (horizontal wall beams attached to the exterior columns in an industrial type building; girts support the exterior wall and provide bending resistance due to wind); *stringers* (beams parallel to the traffic direction in a bridge floor system supported at panel points of trusses located on each side of the bridge deck); and, *diaphrams* (beams in a bridge floor system that span between the girders and provide some wheel load distribution in the direction perpendicular to traffic).

135

5.2 BEHAVIOR DUE TO APPLIED MEMBER END MOMENTS

The most commonly used cross section for a hot rolled steel beam is the W section which is doubly symmetric. A W section is configured for economy to provide much more bending resistance about the major principal axis than about the minor principal axis. Therefore, the following discussion begins with the behavior of a W section beam subjected to an applied moment which causes the member to bend about the major principal axis of the cross section.

Figure 5.1b shows a W section beam of infinitesimal length subjected only to bending about the major principal axis of the cross section. A cross section in the XY plane prior to bending is assumed to remain a plane section in the rotated position after bending occurs. Therefore, as shown in Figure 5.1c, the strain diagram is linear; the *maximum compressive strain* is denoted as ε_c and the *maximum tensile strain* is denoted as ε_t. The rate of change of θ_X, $\dfrac{d\theta_X}{dz}$, in Figure 5.1b is called the *curvature*, ϕ_X, which can be computed as shown in Figure 5.1c for small slope theory.

To simplify the discussion, **temporarily assume** there are not any *residual stresses* in the member. For a particular grade of steel, the *stress-strain relation* for each cross sectional fiber is as shown in Figures 2.1 and 2.2 (see page 2-2.1). If the *extreme fiber strains*, ε_c and ε_t, do not exceed the *yield strain*, ε_y, on the appropriate stress-strain curve, the stress diagram is E times the strain diagram (see the completely elastic case in Figure 5.1d). Alternatively, the bending stresses for the completely elastic case in Figure 5.1d can be computed by using the *flexural formula* :

$$f = \frac{-M_X y}{I_X}$$

where the minus sign accounts for the chosen sign convention: a *tensile stress* is <u>positive</u> and a *compressive stress* is <u>negative</u>. In the cross sectional region where y is negative in Figure 5.1a, only tensile stresses exist in Figure 5.1d due to the applied moment in Figure 5.1b and tensile stresses are positive. Therefore, the minus sign in the flexural formula is needed for the bending stress to have the correct sign.

136

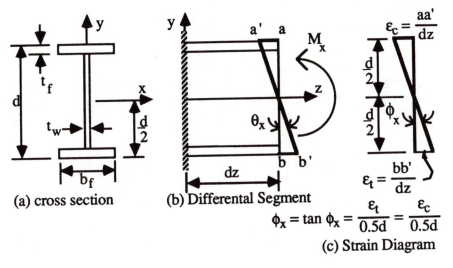

(a) cross section   (b) Differental Segment

$$\phi_x = \tan \phi_x = \frac{\varepsilon_t}{0.5d} = \frac{\varepsilon_c}{0.5d}$$

(c) Strain Diagram

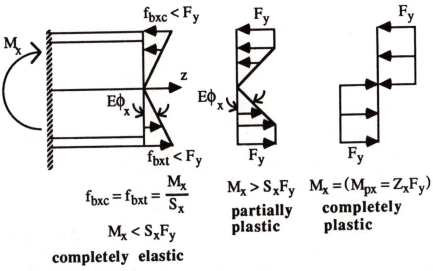

$$f_{bxc} = f_{bxt} = \frac{M_x}{S_x}$$

$$M_x < S_xF_y$$

completely elastic

$M_x > S_xF_y$
partially plastic

$M_x = (M_{px} = Z_xF_y)$
completely plastic

(d) Stress diagrams for various values of M_x

Assumption: There are not any residual stresses in the W section.

Figure 5.1 Uniform bending of a W section of infinitesimal length

In Figure 5.1d, note that the author chose to use arrows instead of signs to denote *compressive* and *tensile stresses* acting on the right end of the beam segment. For the completely elastic case in Figure 5.1d, the extreme fiber stresses due to M_X are denoted as:

f_{bxc} (*maximum* bending compression stress) and
f_{bxt} (*maximum* bending tensile stress).

Eventually we will have to be prepared to deal with cross sections that are not symmetric about the bending axis. Therefore, the author chooses to make the following definitions for a cross section that is not symmetric about the X-axis. Let :

1. y_c and y_t , respectively, be the <u>absolute value of y in the flexural formula</u> when the stress in the *extreme compression fiber* and in the *extreme tension fiber* , respectively, are computed.

2. $S_{xc} = \dfrac{I_X}{y_c}$ and $S_{xt} = \dfrac{I_X}{y_t}$

Then, the *maximum* compression bending stress is: $f_{bxc} = \dfrac{M_X}{S_{xc}}$

and the *maximum* tension bending stress is: $f_{bxt} = \dfrac{M_X}{S_{xt}}$

If the cross section is symmetric about the X-axis, let :

$$S_{xc} = S_{xt} = \left[S_X = \frac{I_X}{(d/2)} \right]$$

$$f_{bxc} = f_{bxc} = \left[f_{bx} = \frac{M_X}{S_X} \right]$$

For each W section(see LRFD page 1-19) the value of S_X , the *elastic section modulus for X-axis bending*, is given in the LRFD Manual.

When M_X in Figure 5.1b produces <u>extreme fiber strains</u> in Figure 5.1c that <u>exceed the yield strain</u> but are less than the strain-hardening strain, the <u>bending stresses</u> are as shown in the partially plastic case of Figure 5.1d and <u>cannot be computed by the flexural formula</u>. **These bending stresses must be obtained from the *stress-strain curve definitions*.** The completely plastic case of Figure 5.1d is theoretically impossible since the strain at the bending axis is always zero. However, in laboratory tests the *plastic bending moment,*

$$M_{px} = Z_X * F_y$$

where Z_X is the *plastic section modulus for X-axis bending*, can be developed and exceeded as shown in Figure 5.2 on the Moment-Curvature diagram. When the curvature in Figure 5.2 is only two times the curvature at first yield, $M_X = 0.97 * M_{px}$. For A36 steel,

$\phi_{sh} \approx 12\phi_{yield}$ (the curvature at which strain hardening <u>begins</u> to occur is about 12 times the yield curvature).

138

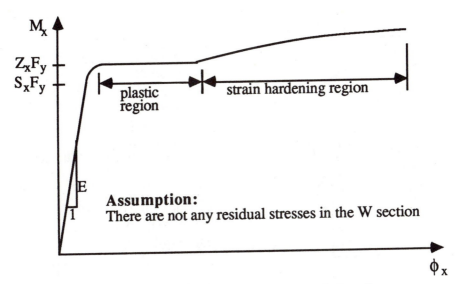

Figure 5.2 Moment-Curvature for uniform x-axis bending
of a W section

As shown in Figure 5.3, there are *residual stresses* in a hot-rolled W section <u>due to uneven cooling</u>. The extreme fiber at each flange tip has the <u>*maximum*</u> *residual compressive stress*, F_{rc}, which is assumed to be 10 ksi in LRFD F1.3(page 6-44). Therefore, based on the assumption stated in the note of Figure 5.3, the top corner fibers in Figure 5.1a begin to yield when :

$$\frac{M_X}{S_X} + F_{rc} = F_y$$

Figure 5.4 shows the effect of residual stresses on the Moment-Curvature relation for X-axis bending of a W section. <u>When residual stresses are accounted for,</u> the elastic limit for M_x is

$$M_{rx} = S_X*(F_y - F_{rc})$$

whereas <u>if residual stresses are ignored</u> the elastic limit for M_x is:

$$M_{(yield)X} = S_X*F_y$$

Also, the fibers that yielded first will be the first fibers to strain harden. However, the bending moment for the completely plastic case of Figure

139

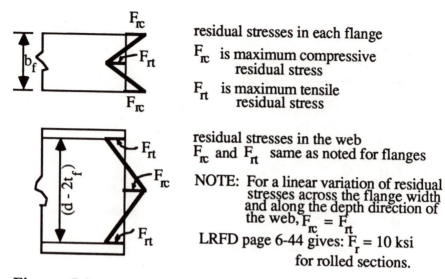

residual stresses in each flange

F_{rc} is maximum compressive residual stress

F_{rt} is maximum tensile residual stress

residual stresses in the web

F_{rc} and F_{rt} same as noted for flanges

NOTE: For a linear variation of residual stresses across the flange width and along the depth direction of the web, $F_{rc} = F_{rt}$

LRFD page 6-44 gives: $F_r = 10$ ksi for rolled sections.

Figure 5.3 Residual stresses in a W section

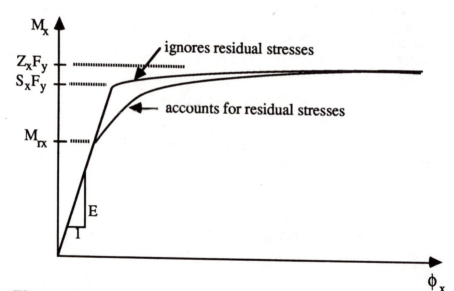

Figure 5.4 Effect of residual stress on M_x vs ϕ_x for a W section

5.1d is adopted as the *maximum nominal bending strength* in the LRFD Specifications.

Figure 5.5a shows a W section beam subjected to applied member end moments that cause bending to occur about the X-axis of the cross section. The bending moment due to the self weight of the member is assumed to be negligible and, as shown in Figure 5.5b, the bending moment is constant along the member length. As shown on LRFD page 1-168, the member has some initial crookedness defined as camber(X-axis bending) and sweep(Y-axis bending).

As shown in Figure 5.6, the member end moments in Figure 5.5a can be replaced with a couple. Due to the tension force of the couple, below the X-axis of the W section the fibers elongate and their sweep crookedness decreases. Due to the compression force of the couple, above the X-axis of the W section the fibers shorten and their sweep crookedness increases. Thus, the compression flange plus a small portion of the adjoining web can be imagined to be a column which will buckle(see Figure 5.6) about the Y-axis of the W section when the axial column force C reaches the critical value. Note that column buckling about the X-axis for the imagined column cannot occur. The bottom half of the member is in tension. Most likely the reader has seen a circus performer walk on a pretensioned wire. Thus, any tendency of the imagined column to move some more in the deflected beam direction(in the negative Y direction) would be transferred through the web of the W section and resisted by the tension force in the bottom half of the member.

If the intermediate lateral braces for the compression flange in Figures 5.5 and 5.6 are removed, we get Figure 5.7 which is the fundamental case in the LRFD definition of *lateral-torsional buckling of a W section subjected to X-axis bending.*

If $L_b = L$ in Figure 5.7 is large enough, elastic lateral-torsional buckling will occur and the critical value of $\bar{M}$ is(see [6,page 253] or [7,page 160]):

$$M_{cr} = \frac{\pi}{L_b}\sqrt{EI_yGJ + \left(\frac{\pi E}{L_b}\right)^2 I_y C_w} \qquad \text{(Eqn 5.1)}$$

which is LRFD Eqn(F1-13)[page 6-44] without the C_b parameter. J is the *torsional constant* of a cross section and C_w is the *warping constant* of a cross section. Values of J and C_w are given on LRFD pages 1-133 to 1-156 for W and T sections, channels, and single angles. It should be noted that $\bar{M} \leq M_{rx}$ (see Figure 5.4); that is, the effect of residual stresses is accounted for in defining the condition of

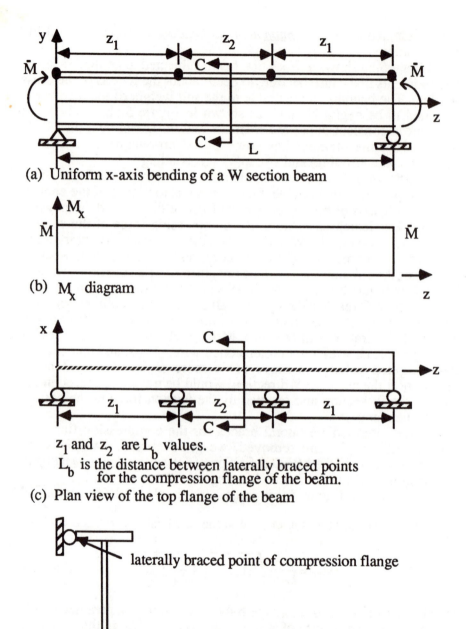

(a) Uniform x-axis bending of a W section beam

(b) M_x diagram

z_1 and z_2 are L_b values.
L_b is the distance between laterally braced points
for the compression flange of the beam.
(c) Plan view of the top flange of the beam

laterally braced point of compression flange

(d) Section C-C

Figure 5.5 W section beam subjected to member end moments

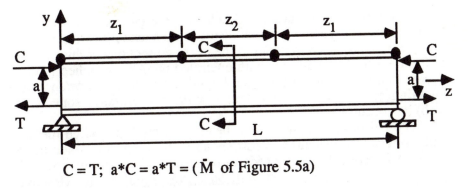

$$C = T; \ a*C = a*T = (\bar{M} \text{ of Figure 5.5a})$$

(a) Conceptually equivalent to Figure 5.5a

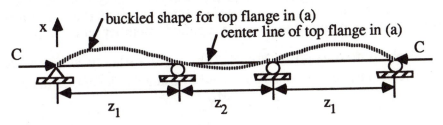

(b) Line diagram plan view of (a)

Figure 5.6 Compression flange of a beam conceptualized as a column

elastic lateral-torsional buckling.

If $L_b = L$ in Figure 5.7 is small enough, the plastic bending moment for X-axis bending, $\mathbf{M_{px}}$, is reached before lateral-torsional buckling occurs and

$$\bar{M} = (M_{px} = Z_X * F_y)$$

is required to produce the described plastic bending moment condition.

Inelastic lateral-torsional buckling occurs before $\bar{M} = M_{px}$ is reached when $L_p < L_b < L_r$
where: $\mathbf{L_b} = L$ in Figure 5.7
$\mathbf{L_r}$ is the smallest value of L_b for which elastic lateral-torsional buckling occurs

$\mathbf{L_p}$ is the largest value of L_b for which $\bar{M} = M_{px}$ is reached before lateral-torsional buckling occurs.

143

A summary of the described behavior for the member in Figure 5.7 is given in Figure 5.8 which also contains some information that the author will discuss in detail in the next two sections of this chapter. In Figure 5.8, the small arcs with encircled numbers attached to them denote that either lateral-torsional buckling has occurred or/and local buckling(of the compression flange or the compression zone of the web) has occurred.

The behavior of a W section beam subjected to applied member end moments which cause uniform bending to occur only about the Y-axis(see Figure 5.1a) also needs to be discussed. For this type of W section bending, *lateral-torsional buckling* cannot occur. However, *local buckling* of each half flange that is in compression due to bending about the Y-axis can occur *unless the width-thickness ratio does not exceed a certain limit* to be specified later(see page 156). **If local buckling is prevented**, as the applied member end moments increase from zero up to $\bar{M} = S_Y*(F_y - 10 \text{ ksi})$, the stresses are completely elastic and can be computed by using the flexural formula:

$$f = \frac{M_Y \, x}{I_Y}$$

and the extreme fiber stresses are:

$$\frac{M_Y}{S_Y} \pm 10 \text{ ksi}$$

where: S_Y is the *elastic section modulus for Y-axis bending* and
the minus sign applies for the flange tips which are in tension due to M_Y

If local buckling is prevented, the *plastic moment for Y-axis bending*, $M_{py} = Z_Y*F_y$ can be reached.

The behavior of other cross sections(channels, tees, double angles) subjected to uniform bending about the major principal axis will be discussed later(see page 156). Also, the behavior of a W section beam subjected to unequal applied member end moments which cause bending to occur about the X-axis will be discussed later(see pages 153-156).

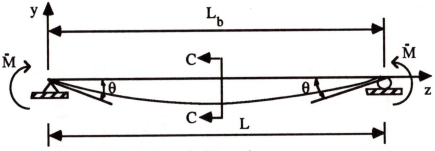

(a) Uniform x-axis bending of a W section

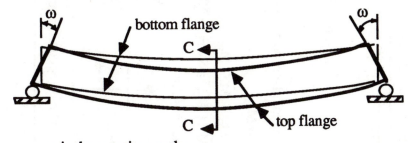

ω is the warping angle.
    (angle between the end planes of the top and bottom flanges)

(b) Plan view of (a)

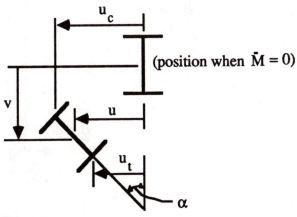

(c) Section C-C

Figure 5.7 Lateral-Torsional buckling of a W section beam

145

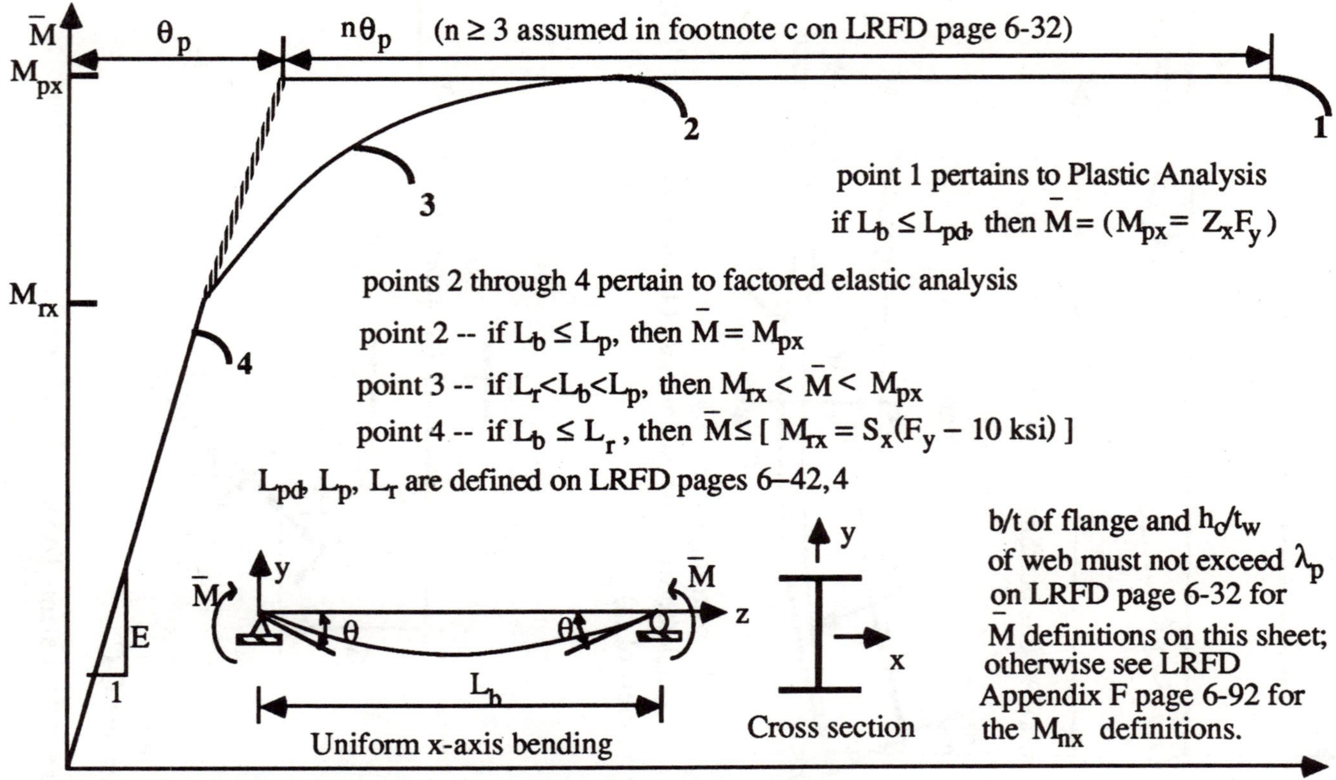
M̄
M_{px}
M_{rx}
θp
nθp (n ≥ 3 assumed in footnote c on LRFD page 6-32)
1
2
3
4
E
1
point 1 pertains to Plastic Analysis
if $L_b \le L_{pd}$, then $\bar{M} = (M_{px} = Z_x F_y)$
points 2 through 4 pertain to factored elastic analysis
point 2 -- if $L_b \le L_p$, then $\bar{M} = M_{px}$
point 3 -- if $L_r < L_b < L_p$, then $M_{rx} < \bar{M} < M_{px}$
point 4 -- if $L_b \le L_r$, then $\bar{M} \le [\, M_{rx} = S_x(F_y - 10\text{ ksi})\,]$
L_{pd}, L_p, L_r are defined on LRFD pages 6–42,4
y
z
θ
θ
M̄
M̄
L_b
Uniform x-axis bending
y
x
Cross section
b/t of flange and h_c/t_w
of web must not exceed λ_p
on LRFD page 6-32 for
M̄ definitions on this sheet;
otherwise see LRFD
Appendix F page 6-92 for
the M_{nx} definitions.
θ

Figure 5.8 Member-end M vs θ for x-axis bending of a W section with (b/t of flange and h_c/t_w) ≤ (λ_p on LRFD page 6-32)

5.3 LIMITING WIDTH-THICKNESS RATIOS FOR COMPRESSION ELEMENTS

When a W section beam is subjected to bending about the X- axis, one flange and half of the web are in compression(see Figure 5.1 page 137). Therefore, _local buckling_ of these compression elements can occur and can be the phenomenon that controls the design bending strength if the W section elements are not properly configured. As shown in Figure 5.7(page 145), _lateral-torsional buckling_ of the beam can also occur and can be the phenomenon that controls the design bending strength.

Local buckling and _lateral-torsional buckling_ are not always independent phenomena. For discussion purposes, suppose we choose a W10X26 section of A36 steel for some beam tests to be conducted in a laboratory. In each beam test described below, local buckling of the web cannot occur(the reader can verify this after the author has completed the discussion on the limiting width-thickness ratios for compression elements). In each beam test, an identical W10X26 section of length L is subjected to uniform X-axis bending and _the only variable is_ L_b (**the distance between the laterally braced points of the compression flange**).

In the first beam test, the compression flange is laterally braced at:

$$L_b = [L_p = \frac{300r_y}{\sqrt{F_y}}]$$

where L_p is the largest possible value of L_b for which M_{px} can be reached. The compression flange will begin to deflect noticeable in the lateral direction when M_{px} is reached at about $\theta_x = 2\theta_p$ in Figure 5.8, but the capacity to resist moment will not be reduced until _local buckling_ in the compression flange occurs at $\theta_x \geq 3\theta_p$.

In the second beam test, $L_b = 0.5*L_p$ and the beam length, L, is the same as in the first beam test. M_{px} will be reached as before, but the capacity to resist moment will not be reduced until one of the following conditions occurs at $\theta_x \geq 3\theta_p$:

1) _local buckling_ of the compression flange will occur and either immediately or after a very short time lapse _lateral- torsional buckling_ will occur.
2) _lateral-torsional buckling_ will occur and either immediately or after a very short time lapse _local buckling_ of the compression flange will occur.

147

In the third beam test , $L_b = 0$(the compression flange is continuously laterally braced along the beam length, L, which is the same as in the previous tests). M_{px} will be reached and *local buckling* of the compression flange will occur at about $\theta_x = 9\theta_p$, but lateral-torsional buckling cannot occur.

The preceding discussion indicates that local buckling and lateral-torsional buckling are not necessarily independent phenomena. For simplicity, wherever it is possible to do so, they are treated independently in the research literature and in the theoretically oriented textbooks[6,7]. At the first glance of the LRFD Specifications(see LRFD pages 6-31,32), it appears that local buckling is treated independently. However, upon looking more carefully we find that items 2 and 3 at the top of LRFD page 6-33 are listed under the topic entitled LOCAL BUCKLING. The author teaches a graduate level course devotely entirely to Plastic Analysis and the LRFD Specifications cited in B5.2 on LRFD page 6-33, but he does not teach Plastic Analysis to his undergraduate students. The most complicated (less frequently needed) LRFD Specifications are located in the LRFD Appendices. Therefore, it is not appropriate to take an in depth look at LRFD Appendix F1.7(page 6-92) cited in LRFD B5.3 (page 6-33) at this point in this textbook. Consequently, the reader will have to accept the author's statement that local buckling and lateral-torsional buckling are not treated as two completely independent topics in the LRFD Specifications. Furthermore, the author believes that students will gain a better understanding of the material which needs to be presented now if he shows how lateral-torsional buckling is related to local buckling in the AISC Specifications.

For a W section or a channel subjected to bending about the X-axis to reach M_{px} (see item 2 on Figure 5.8) and to develop an inelastic rotation of at least $3\theta_p$ before either local buckling or lateral-torsional buckling occurs, the following requirements **must be satisfied**:
1) for the flanges(see first item on LRFD page 6-32):

$$\frac{b}{t} \leq \frac{65}{\sqrt{F_y}}$$

For a W section(see LRFD page 6-31 *item a of an unstiffened element*):

$$\frac{b}{t} = \frac{0.5b_f}{t_f}$$

NOTE: $\dfrac{0.5b_f}{t_f}$ is listed for each W section(see LRFD page 1-37).

For a channel(see LRFD page 6-31 *item b of an unstiffened element*):
$$\frac{b}{t} = \frac{b_f}{\text{average } t_f} \quad \text{[see LRFD page 1-44]}$$

2) for the web(see LRFD page 6-32, webs in flexural compression):
$$\frac{h_c}{t_w} \le \frac{640}{\sqrt{F_y}}$$

See LRFD page 6-31 *item a of a stiffened element* for the definition of h_c. $\dfrac{h_c}{t_w}$ is listed for each W section(see LRFD page 1-37). For a channel(see LRFD page 1-44), $h_c \approx T$.

3) for the compression flange(see LRFD page 6-43):
$$\text{each } L_b \le [L_p = \frac{300r_y}{\sqrt{F_y}}]$$

where L_b is the *distance between two adjacent laterally braced points* .

As the author has shown, the limiting width-thickness ratios for compression elements(see LRFD page 6-32) in a beam cross section and the unbraced length of the compression flange of a beam are parameters which must be used in the determination of the nominal bending strength. Therefore, the author chooses to locate further discussions of the limiting width-thickness ratios for compression elements of beam cross sections in the topic entitled Design Bending Strength.

5.4   DESIGN BENDING STRENGTH

If the *width-thickness ratio* of each compression element in either a singly or doubly symmetric cross section of a beam does not exceed the applicable λ_p value given on LRFD page 6-32, the LRFD definition of the *X-axis flexural design strength* is $\phi_b M_{nx}$ where $\phi_b = 0.9$ and M_{nx} is the *X-axis nominal bending strength*.

149

See LRFD F1.2(page 6-42) for the preceding definition of the *flexural design strength*. The author prefers to say this is the definition of the **X-axis design bending strength** to be consistent with previously introduced LRFD terminologies: <u>design tensile strength</u> (for tension members), and <u>design compressive strength</u>(for columns).

The *X-axis nominal bending strength* <u>for</u> a beam whose cross section is either a <u>W section</u> or a <u>channel</u>(C section) is:

1) If $\dfrac{0.5b_f}{t_f} \le (\lambda_p = \dfrac{65}{\sqrt{F_y}}), \ \dfrac{h_c}{t_w} \le \left[\lambda_p = \dfrac{640}{\sqrt{F_y}}\right],$ and $L_b \le \left[L_p = \dfrac{300r_y}{\sqrt{F_y}}\right]$

$$M_{nx} = (M_{px} = Z_X * F_y) \qquad \text{(Eqn 5.2)}$$

2) If $\dfrac{0.5b_f}{t_f} \le (\lambda_p = \dfrac{65}{\sqrt{F_y}}), \ \dfrac{h_c}{t_w} \le \left[\lambda_p = \dfrac{640}{\sqrt{F_y}}\right],$ and $L_p < L_b \le L_r$

where: $L_r = \dfrac{r_y X_1}{F_y - 10 \text{ ksi}} \sqrt{1 + \sqrt{1 + X_2(F_y - 10 \text{ ksi})^2}}$

$$X_1 = \dfrac{\pi}{S_X}\sqrt{\dfrac{EGJA}{2}}$$

$$X_2 = \dfrac{4C_w}{I_y}\left(\dfrac{S_X}{GJ}\right)^2$$

$$M_{nx} = C_b \left[M_{px} - (M_{px} - M_{rx})\left(\dfrac{L_b - L_p}{L_r - L_p}\right)\right] \le M_{px} \qquad \text{(Eqn 5.3)}$$

where: $M_{rx} = S_X(F_y - 10 \text{ ksi})$

C_b is a parameter which accounts for the case where the M_X diagram is not constant in an L_b interval.

$C_b = 1.0$ for each L_b interval in which the moment is greater than or equal to M_2 and for a cantilever beam

otherwise $C_b = 1.75 + 1.05(M_1/M_2) + 0.3(M_1/M_2)^2$

where: M_1 and M_2 are the moments at the ends of an L_b interval

M_1 is the smaller end moment

M_2 is the larger end moment

M_1/M_2 is <u>positive</u> when the end moments cause <u>reverse curvature</u> to occur as shown below:

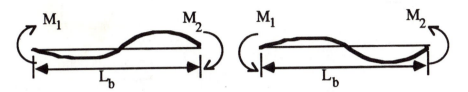

M_1/M_2 is <u>negative</u> when the end moments cause <u>single curvature</u> to occur as shown below:

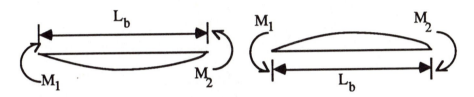

3) If $\dfrac{0.5b_f}{t_f} \leq (\lambda_p = \dfrac{65}{\sqrt{F_y}})$, $\dfrac{h_c}{t_w} \leq \left[\lambda_p = \dfrac{640}{\sqrt{F_y}}\right]$, and $L_b > L_r$

$$M_{nx} = M_{cr} \leq C_b M_r \qquad \text{(Eqn 5.4)}$$

where: $M_{cr} = \dfrac{\pi}{L_b}\sqrt{EI_yGJ + \left(\dfrac{\pi E}{L_b}\right)^2 I_y C_w}$

NOTE: M_{nx} vs L_b defined in items 1, 2 and 3 above is conceptually illustrated in Figure 5.9 for $C_b = 1$ in items (2) and (3) above. See Figures 5.10 to 5.12 <u>for the effect of $C_b > 1$</u> on the *nominal bending strength.*

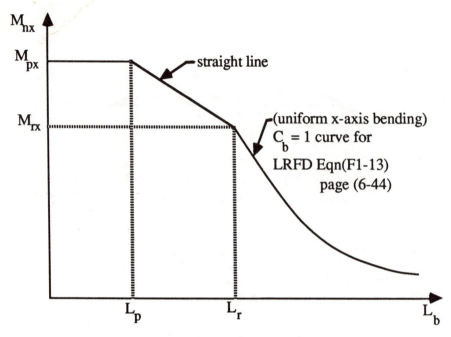

NOTE: b/t of flange and h_c/t_w of web
must not exceed λ_p given on LRFD page 6-32.

Figure 5.9 M_{nx} vs L_b for a W section subjected to x-axis bending

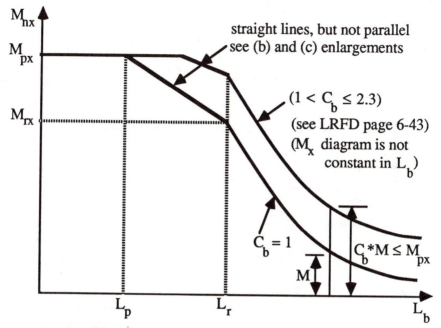

NOTE: b/t of flange and h_c/t_w of web
must not exceed λ_p given on LRFD page 6-32.

(a) An example for $(1 < C_b \leq 2.3)$

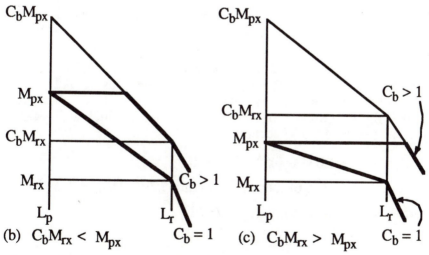

(b) $C_b M_{rx} < M_{px}$

(c) $C_b M_{rx} > M_{px}$

Figure 5.10 M_{nx} vs L_b for a W section when $C_b > 1$

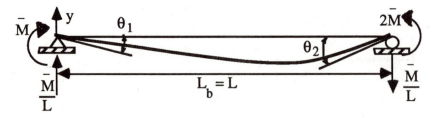

(a) Single curvature x-axis bending of a W section beam

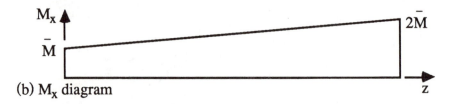

(b) M_x diagram

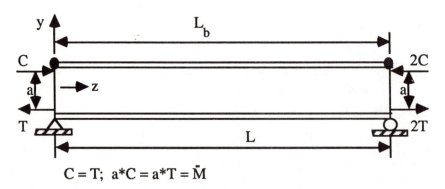

$$C = T; \quad a*C = a*T = \bar{M}$$

Note that the compressive force in the top half of the beam linearly increases from C to 2C. The compression flanges plus a small portion of the akjoing web can be imagined to be a column subjected to C + (C/L uniformly distributed along the length) which will buckle about the y-axis of the W section.

$$C_b = 1.75 + 1.05*(-0.5) + 0.3*(-0.5)^2 = 1.3$$

$$M_{crx} = [C_b*(M_{crx} \text{ of member in Figure 5.7})] \leq [M_{px} = Z_x F_y]$$

$$(M_{crx} \text{ of member in Figure 5.7}) = \bar{M}$$

(c) Conceptually equivalent to (a)

Figure 5.11   Effect of $C_b > 1$ on Lateral–Torsional Buckling

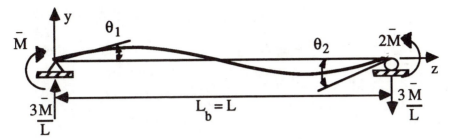

(a) Single curvature x-axis bending of a W section beam

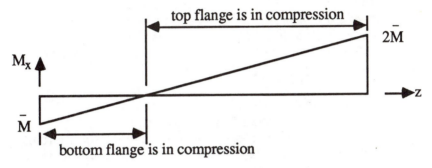

(b) M_x diagram

$$C_b = 1.75 + 1.05*(+0.5) + 0.3*(+0.5)^2 = 2.35$$

LRFD Specs. limit C_b to 2.3

$$M_{crx} = [C_b*(M_{crx} \text{ of member in Figure 5.7})] \le [M_{px} = Z_x F_y]$$

$(M_{crx} \text{ of member in Figure 5.7}) = \overline{M}$

Note that the compressive force in the top flange varies linearly from a maximum value at the right end to zero at the $M = 0$ point. Likewise, the compression force in the bottom flange varies from a maximum value at the left end to the $M = 0$ point.

(c) Comments

Figure 5.12 Reverse bending effect on Lateral-Torsional Buckling

If either $\frac{0.5b_f}{t_f}$ or $\frac{h_c}{t_w}$ of a W section or a channel subjected to X-axis bending exceeds the value of λ_p (see λ_p formulas shown above and on LRFD page 6-32) but does not exceed the value of λ_r (see formulas shown on LRFD page 6-32), LRFD Appendix F(page 6-92) must be used to determine the *nominal bending strength* which is controlled either by *inelastic local buckling* or by *inelastic lateral-torsional buckling*.

If either $\frac{0.5b_f}{t_f}$ or $\frac{h_c}{t_w}$ of a W section or a channel subjected to X-axis bending exceeds the value of λ_r (see formulas shown on LRFD page 6-32), LRFD Appendix F(page 6-92) must be used to determine the *nominal bending strength* which is controlled either by *elastic local buckling* or by *elastic lateral-torsional buckling*.

The *Y-axis nominal bending strength* of a beam whose cross section is either a W section or a channel is:

If $\frac{0.5b_f}{t_f} \le (\lambda_p = \frac{65}{\sqrt{F_y}})$, $\quad M_{ny} = (M_{py} = Z_Y * F_y)$ $\qquad$ (Eqn 5.7)

otherwise see LRFD Appendix F1.7(page 6-92).

Eqn(5.7) also applies for any bending axis of solid circular and square sections as well as for bending about the minor principal axis of any non-built-up section[see LRFD F1.7(page 6-45)]. For minor axis bending, if the width-thickness ratio of any compression element exceeds the appropriate λ_p value shown on LRFD page 6-32, LRFD Appendix F(page 6-92) must be used to determine the *nominal bending strength*.

For a beam whose cross section is either a tee section (see LRFD pages 1-62 to 1-81) or a double angle section (see LRFD pages 1-84 to 1-95) with zero separation between the back to back legs, the *X-axis nominal bending strength* is:

If $\frac{0.5b_f}{t_f} \le (\lambda_r = \frac{95}{\sqrt{F_y}})$ and $\frac{d}{t_w} \le \left[\lambda_p = \frac{127}{\sqrt{F_y}}\right]$

$$M_{nx} = M_{cr} \le (M_y = S_X F_y) \qquad \text{(Eqn 5.5)}$$

$$M_{cr} = \frac{C_b \pi}{L_b} \sqrt{EI_y GJ} \left[B + \sqrt{1 + B^2} \right]$$

$$B = \pm 2.3 \frac{d}{L_b} \sqrt{\frac{I_y}{J}} \qquad \text{(Eqn 5.6)}$$

where, for B, the plus sign applies when the web part below the X-axis (see figures on LRFD pages 1-62 and 1-88 with 0.375 in. separation deleted) is in tension and the minus sign applies when the web part below the X-axis is in compression due to bending about the X-axis.

The *X-axis nominal bending strength* for a beam whose cross section is a double angle section with connector plates between the back to back legs at intervals along the member length **is as defined in Eqns(5.5,6),** <u>provided</u> $\frac{b}{t} \leq \left[\lambda_r = \frac{76}{\sqrt{F_y}}\right]$ for each leg in each angle.

5.5 HOLES IN BEAM FLANGES

See LRFD B1(page 6-29) which states: "Plate girders, coverplated beams and rolled or welded beams shall be proportioned on the basis of the gross section. No deduction shall be made for shop or field bolt holes in either flange unless the reduction of the area of either flange by such holes, calculated in accordance with the provisions of Sect. B2, exceeds 15% of the gross flange area, in which case the area in excess of 15% shall be deducted."

5.6 DEFLECTIONS

In the second paragraph on LRFD page 6-81, we find: "Limiting values of structural behavior to ensure serviceability (e.g., maximum deflections, accelerations, etc.) shall be chosen with regard to the intended function of the structure."

LRFD L3.1(page 6-81) states that: "Deformations in structural members and structural systems due to service loads shall not impair the serviceability of the structure."

LRFD L3.3(page 6-81) states that: "Lateral deflection or drift of structures due to code-specified wind or seismic loads shall not cause collision with adjacent structures nor exceed the limiting values of such drifts which may be specified or appropriate."

The LRFD Specifications do not provide any guidelines on the limiting values for beam deflections and for drift(system deflections due to and in the direction of wind, for example). Therefore, the structural designer must decide what the appropriate limiting deflection values for each structure are based on experience, judgement, the satisfactory performance of a similar structure, and the owner's intended usage of the structure. After the author has given a general discussion on the need for controlling deflections, he will give some suggested limiting values for beam and drift deflections.

Deflections must be considered in the design of almost every structure. In the interest of minimizing the dead weight of high rise structures, high strength steel members are used wherever they are economically and structurally feasible. Consider, for example, a W section member which is 20 ft. long. The member weight is a function of only one variable, the cross sectional area. As the member weight decreases, the cross sectional area and the moments of inertia decrease, the member becomes more flexible and permits larger deflections to occur. Consequently, controlling deflections becomes more of a problem when the dead weight of the steel members is minimized. If a high rise structure sways too much or too rapidly, the occupants become nauseated or frightened at best although no structural damage may occur. Similarly, the public becomes alarmed if a floor system of a building or a bridge is too flexible and noticeably sags more than a tolerable amount. Also there are situations, such as a beam over a plate glass window or a water pipe, where excessive deflections can cause considerable damage if they are not controlled by the structural designer. If the roof beams in a flat roof sag too much, water ponds on the roof causing additional sagging, more ponding which can rupture the roof surface, and extensive water damage can occur to the contents of the structure. Consequently, it is not unusual for deflections to be the controlling factor in the design of a structure or the design of a member in a structure.

Construction can only be done within tolerable limits -- for example, columns cannot be perfectly plumbed and foundations cannot be placed perfectly in plan view nor in elevation view. Deflections that occur during construction due to these imperfect erection conditions, wind, temperature changes, and construction loads must be controlled by the steel erection contractor to ensure the safety of the structure, the construction workers, and the public. Total collapses of steel structures have occurred during construction because the erection contractors did not provide adequate drift control bracing or/and adequate shoring to limit gravity direction deflections during construction.

Deflections due to temperature changes may need to be controlled. The St. Louis arch memorial was constructed by cantilevering independently from the two foundations toward the crown of the arch. The last prefabricated segment of the arch was inserted between the two independently erected cantilevered parts of the arch. Since the arch lies in the North-South plane, the top of the southern half of the arch was

158

directly exposed to the sun at high noon whereas the top of the northern half of the arch was shaded (due to the curvature of the arch). At high noon the differential displacements at the tips of the two independently erected cantilevers were approaching their maximum amounts. The contractor preferred to insert the last segment at the crown shortly after daybreak when the temperature of the two cantilevers were nearly the same. However, for maximum public relations purposes, public officials decided to have the last segment of the arch inserted at high noon and arranged for the St. Louis Fire Department to hose down the southern half of the arch with water to minimize the differential deflections caused at the cantilevered tips by the two different sun exposures.

The deflected shape of a structure due to service conditions sometimes can be economically controlled by pre-cambering the structure. Pre-cambering[see LRFD L1(page 6-810] is achieved by erecting the structure with built-in deformations such that the structure deflects to or slightly below its theoretical no-load shape when the maximum service loads occur on the structure. For example, suppose the bottom chords of a simply supported plane truss are deliberately fabricated too short. When the truss is assembled, the interior truss joints displace upwardly. When the service loads occur, each interior truss joint displaces downwardly to or slightly below its no-load, camberless position.

Current structural design practice is to design the structural members to have adequate strength to resist bending moment, shear, and axial force(if applicable). Then, gravity direction and drift deflections are checked to determine if they are adequately controlled to ensure the desired level of serviceability. Some of the common servicability problems are[3]:

1. Local damage of nonstructural elements (for example, windows, ceilings, partitions, walls) occurs due to displacements caused by loads, temperature changes, moisture, shrinkage, and creep.
2. Equipment (for example, elevators) does not function normally due to excessive displacements.
3. Drift or/and gravity direction deflections are so noticeable that occupants become alarmed.
4. Extensive nonstructural damage occurs due to a tornado or a hurricane.
5. Structural deterioration occurs due to age and usage (for example, deterioration of bridges and parking decks due to de- icing salt).

159

6. Motion sickness of the occupants occurs due to excessive floor vibrations caused by routine occupant activities or lateral vibrations due to the effects of wind or an earthquake.

These serviceability problems can be categorized as a function of either the gravity direction deflection or the lateral deflection.
Let: L = span length of a floor or roof member; h = story height.

Deflection Index	Typical Serviceability Behavior
h/1000	Not visible cracking of brickwork
h/500	Not visible cracking of partition walls
h/300 or L/300	Visible architectural damage Visible cracks in reinforced walls Visible ceiling and floor damage Leaks in the structural facade
L/200 to L/300 h/200 to h/300	Cracks are visually annoying Visible damage to partitions and large, plate glass windows
L/100 to L/200 h/100 to h/200	Visible damage to structural finishes Doors, windows, sliding partitions, and elevators do not function properly

It is customary steel design practice to limit the Deflection Index to:
1. L/360 **due to live load** on a floor or snow load on a roof when the beam supports a plastered ceiling;
2. L/240 **due to live load or snow load** if the ceiling is not plastered;
3. h/667 to h/200 for each story due to the effects of wind or earthquakes -- only a range of limiting values can be given for many reasons (type of facade, activity of the occupants, routine design, innovative design, structural designer's judgment and experience);
4. H/715 to H/250 for entire building height H due to the effects of wind or earthquakes -- comment in item 3 applies here too.

It is important to note that: the Deflection Index limits for drift are about the same as the accuracy which can be achieved in the erection of the structure; and, the largest tolerable deflection due to live load is

160

0.5% of the member length. Consequently, deflections are grossly exaggerated for clarity on deflected structure sketches in textbooks.

5.7   DESIGN SHEAR STRENGTH

The *design requirement for shear* is: $\phi_v V_n \geq V_u$
where V_u is the *required shear strength* (maximum ordinate on the factored shear diagram), $\phi_v V_n = 0.90 * V_n$ is the *design shear strength*, and V_n is the *nominal shear strength* which is defined below.

The design approach for a W section beam, for example, is to select the lightest section that satisfies the design requirement for bending and to check the selected section for all of the other design requirements[shear, serviceability (deflection and vibration control), and others which the author has not yet mentioned]. If any of the other design requirements are not satisfied for the selected section, the structural designer must either choose another section which satisfies all of the design requirements or appropriately modify the selected section to satisfy all of the violated design requirements.

For W section and C section beams, satisfying the design requirement for shear usually is not a problem, except for the following cases:

1. a beam end is coped(see Figure C-J5.2 on LRFD page 6-187).
    If the beam end has the bolted connection depicted in Figure C-J5.2, the *design shear rupture strength* [see LRFD J4(page 6-72), J5.2 item c and the subsequent paragraph on LRFD page 6-74] must be determined. If the beam end is coped but does not have the bolted connection depicted in the cited figure, the design shear rupture strength is not applicable but the coped web depth must be used in the nominal shear strength definition[LRFD F2.2(page 6-45)].

2. holes are made in the beam web for electrical, heating, and air conditioning ducts, for example. The net web depth must be used in the nominal shear strength definition. Also, web stiffeners may be needed around the holes to strengthen the web.

3. the beam is subjected to a large concentrated load located near a support. If **a/d < 2** for *X-axis W section bending*, *shear* probably will control the design; **d** is the *depth* of the W section; **a** is the *distance from the support to the concentrated load nearest the support*. Most likely, *web stiffeners* [see LRFD K1.8(page 6-79) will be needed at the support and at the concentrated load points, and the author would satisfy LRFD G3(page 6-103) and G5(page

161

6-104) in the region between the web stiffeners. If **a/b_f < 0.75** *for Y-axis W section bending,* shear <u>probably will control the design;</u> **b_f** is the *flange width* ; **a** is the *distance from the support to the concentrated load nearest the support.* Each flange must resist half of the shear.

For <u>X-axis bending</u> of W and C sections without any holes in the web and if the beam end is not coped, the *nominal shear strength definition is*[see LRFD F2.2(page 6-45)]:

1) if $\dfrac{h_c}{t_w} \le \dfrac{418}{\sqrt{F_y}}$, *shear yielding of the web* occurs at

$\dfrac{F_y}{\sqrt{3}} = 0.577F_y \approx 0.6F_y$, only the web resists shear, the entire web yields, and $\mathbf{V_n = 0.6F_y t_w d}$

2) if $\dfrac{418}{\sqrt{F_y}} < \dfrac{h_c}{t_w} \le \dfrac{523}{\sqrt{F_y}}$, *inelastic shear buckling of the web*

occurs and $\mathbf{V_n = 0.6F_y t_w d *\dfrac{418/\sqrt{F_y}}{h_c/t_w}}$

3) if $\dfrac{h_c}{t_w} > \dfrac{523}{\sqrt{F_y}}$, *elastic shear buckling of the web* occurs and

$\mathbf{V_n = 132{,}000\dfrac{t_w d}{(h_c/t_w)^2}}$

The LRFD Specifications do not give a shear design strength definition for <u>Y-axis bending</u> of W and C sections.

For other cross sections for which a design shear strength definition is given, see LRFD F2(page 6-45); for plate girders see G(page 6-47) and Appendix G(page 6-101).

Also, see Figure C-K1.2 on LRFD page 6-191 for a sketch pertaining to LRFD K1.7(page 6-79) which gives the design shear strength definition for a column web panel subjected to high shear.

5.8 WEB AND FLANGES SUBJECTED TO CONCENTRATED LOADS

Most likely, the reader has heard the statement "a picture is worth a thousand words". The author believes that the best way to present the material for the indicated topic is to show the applicable LRFD

Specifications on each figure given to aid in the understanding of the presented material.

In Figure 5.13, a tension force is applied perpendicular to a W section beam flange. For clarity, the author chose to show a single plate welded to the flange. However, the single plate could be one of the flanges of another W section whose member end is welded to the bottom flange of the W section shown. In that case, there would be two plates and two T_u forces. The web of the attached W section would also be welded to the bottom flange of the W section shown. The tension force from the attached web would not cause any local bending of the existing W section flange since the webs of both W sections would be in the same plane. However, the tension force from the attached web would have to be accounted for in checking the local web yielding design requirement[LRFD K1.3(page 6-77)]. The stress distribution in the web due to T_u is assumed to be uniform on the section at **k** from the surface of the flange and acting on a **web area** = t_w*L_w where $L_w = 5k + t$ and **t** is the *thickness of the attached plate*. If LRFD K1.2(page 6-77) is not satisfied, a pair of web stiffeners will have to be designed as a tension member to provide the excess amount of the tension force. Also, if LRFD K1.3 is not satisfied, a pair of web stiffeners will have to be designed as a tension member to provide the excess amount of the tension force. If both LRFD K1.2 and K1.3 are not satisfied, a pair of web stiffeners will have to be designed as a tension member to provide the sum of the excess tension force from the indicated specifications.

In Figure 5.14, both flanges of a column W section are subjected to concentrated tension forces and concentrated compression forces. The author has indicated the applicable LRFD Specifications which must be satisfied. A pair of column web stiffeners will have to be designed for each region in which any of the indicated LRFD Specifications is not satisfied. If more than one of these indicated LRFD Specifications is not satisfied, the pair of column web stiffeners will have to be designed as either a tension member(at the top beam flanges in Figure 5.14a) or as a compression member[at the bottom beam flanges in Figure 5.14a -- see LRFD K1.8(page 6-79)] to resist the sum of the excess forces for the violated LRFD Specifications.

163

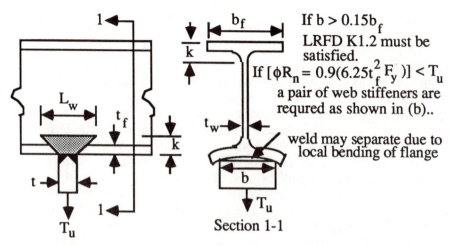

If $b > 0.15b_f$
LRFD K1.2 must be
satisfied.
If $[\phi R_n = 0.9(6.25t_f^2 F_y)] < T_u$
a pair of web stiffeners are
requred as shown in (b)..

weld may separate due to
local bending of flange

Section 1-1

L_w is length along web in which local web yielding may occur

$L_w = 5k + t$ [LRFD K1.3 page 6-77]

If $[\phi R_n = 1.0(5k + t)t_w F_y] < T_u$, a pair of web stiffeners
is required as shown in (b).

(a) Example of local flange bending [LRFD K1.2 page 6-77]

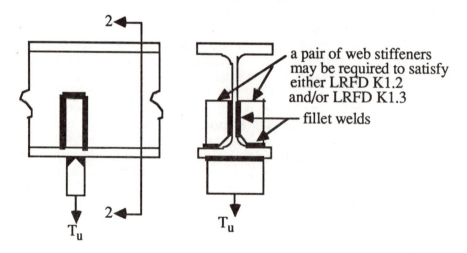

a pair of web stiffeners
may be required to satisfy
either LRFD K1.2
and/or LRFD K1.3

fillet welds

(b) Web stiffeners to prevent either excessive local bending of flange
and/or excessive yielding of web

Figure 5.13   Tension force perpendicular to a W section flange

164

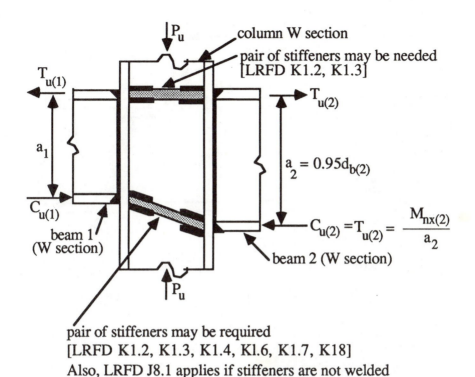

$T_{u(1)}$

a_1

$C_{u(1)}$

beam 1
(W section)

P_u

column W section

pair of stiffeners may be needed
[LRFD K1.2, K1.3]

$T_{u(2)}$

$a_2 = 0.95d_{b(2)}$

$C_{u(2)} = T_{u(2)} = \dfrac{M_{nx(2)}}{a_2}$

beam 2 (W section)

P_u

pair of stiffeners may be required
[LRFD K1.2, K1.3, K1.4, Kl.6, K1.7, K18]
Also, LRFD J8.1 applies if stiffeners are not welded
to the column flanges.

(a) Beam flanges welded to column flanges

See Figure C-K1.2 on LRFD page 6-191 for a set of beam and
column forces that exist due to wind plus gravity loads.

(b) Another example of beam flanges welded to column flanges

Figure 5.14   Tension and compression forces applied
perpendicular to flanges of a column W section

165

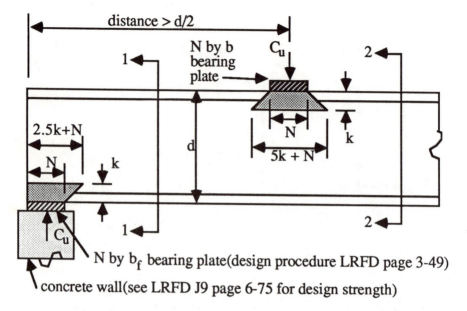

LRFD K1.2, K1.3[Eqn(K1-3)], K1.4 apply at the wall support
If any of the $\phi R_n < C_u$, a pair of web stiffeners is required(Sect. 1-1)
Top flange of W section must be laterally braced at the supports.

LRFD K1.2, K1.3[Eqn(K1-2)], K1.4, K1.5 apply at the interior load
If any of the $\phi R_n < C_u$, a pair of web stiffeners is required(Sect. 2-2)

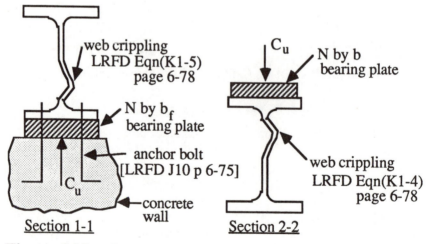

Figure 5.15 Compression force perpendicular to beam flanges

In Figure 5.15, a W section beam is simply supported at the member ends on a wall. A steel bearing plate must be designed at Section 1-1 to spread the beam reaction uniformly on:
1. the wall beneath the bearing plate[see LRFD J9(page 6-75)].
    If the bearing plate is not provided, the
    bearing area = $2k_1$*(length of the beam in contact with the wall)
    See LRFD page 1-36 for the definition of k_1.
2. the web at the toe of the fillet of the W section beam
    [see LRFD K1.3(page 6-77) item b, K1.4(page 6-78)]
See LRFD page 3-49 for the AISC recommended procedure.
The top flange of the beam must be laterally supported at the member ends(see Section 5.9 below). If some means of lateral bracing is not otherwise provided, either an end plate must be welded on the W section beam end or web stiffeners must be welded to the web and both flanges to provide lateral bracing for the top flange of the beam at the beam ends.
    In Figure 5.15 at Section 2-2, the bearing plate is provided to spread the load uniformly on the web at the toe of the fillet of the W section beam. There is not any AISC recommended procedure for the design of this bearing plate. Alternatively, bearing stiffeners(see LRFD K1.8 page 6-79) can be designed instead of a bearing plate.

5.9 LATERAL SUPPORT

    Figures 5.16 and 5.17 show examples of how the compression flange of a W section beam can be laterally braced. The top W section flange in these figures is assumed to be the compression flange. It must be noted that twisting of the beam cross sections at the beam ends(at the beam supports for gravity direction loads) must be prevented. If an interior lateral brace does not prevent twisting of the cross section, LRFD K1.5(page 6-78) must be satisfied.
    Continuous lateral support($L_b = 0$) is provided for the top beam flange in Figure 5.16a if the concrete slab is attached to supports which prevent translation of the slab in the direction perpendicular to the web of the W section beam. See LRFD Figure C-I3.3(page 6-173) which shows shear studs welded to the top flange of a W section at specified intervals along the length direction of the W section. If the specified interval between the shear studs is small enough, continuous lateral support is provided for the top beam flange. Also, the shear studs prevent the concrete slab from slipping along the length direction of the beam and a composite section is formed. A portion of the concrete slab is the top of the composite section and the W section is the remainder of

the composite section. Note the various cross sections of metal decking shown on LRFD page 6-173. If the shear studs on LRFD page 6-173 are omitted and if the metal decking is adequately welded to the top beam flange at specified intervals along the beam length, continuous lateral support can be provided for the top flange of the beam.

As shown in Figure 5.16b, X-braces can be used to provide lateral bracing at intervals along the beam length. The cross braces can be single angles, for example. Alternatively, as shown in Figures 5.16c to 5.16f, cross beams can be used to provide lateral bracing at intervals along the beam length. As shown in Figure 5.17a, a truss must be provided to prevent translation of the cross beams and the X-braces in the direction perpendicular to the web of the W section being laterally braced.

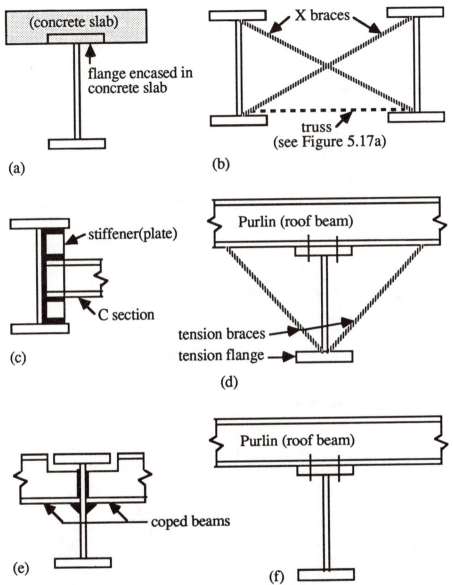

(a)

(concrete slab)

flange encased in
concrete slab

(b)

X braces

truss
(see Figure 5.17a)

(c)

stiffener(plate)

C section

(d)

Purlin (roof beam)

tension braces
tension flange

(e)

coped beams

(f)

Purlin (roof beam)

W section shown in cross section in each figure above is the member
being laterally braced. LRFD K1.5(page 6-78) must be satisfied for
(a), (e), and (f).
NOTE: Also see Figure 5.17

Figure 5.16 Examples of lateral support for a W section beam

169

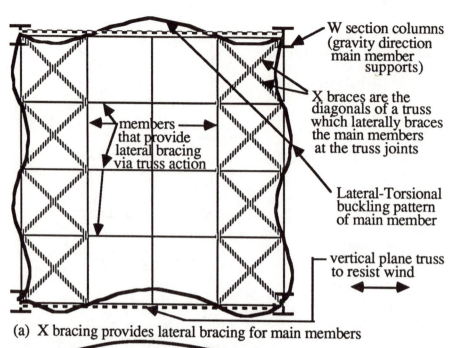

W section columns
(gravity direction
main member
supports)

X braces are the
diagonals of a truss
which laterally braces
the main members
at the truss joints

members
that provide
lateral bracing
via truss action

Lateral-Torsional
buckling pattern
of main member

vertical plane truss
to resist wind

(a) X bracing provides lateral bracing for main members

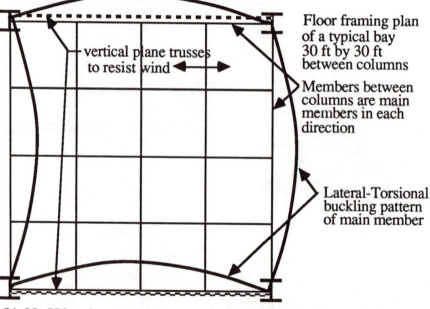

Floor framing plan
of a typical bay
30 ft by 30 ft
between columns

vertical plane trusses
to resist wind

Members between
columns are main
members in each
direction

Lateral-Torsional
buckling pattern
of main member

(b) No X bracing: main members laterally braced only at columns

Figure 5.17 Lateral bracing of main beams in a floor system

170

5.10 EXAMPLE PROBLEMS --
X-AXIS BENDING OF W SECTIONS

In each of the example problems, the objectives are to:
1. show how the beam design aids in the LRFD Manual can be used to select W section beams
2. illustrate the applicable LRFD Specifications

In order to enable the reader to quickly locate an example for a specific type of problem, an index summary description of the example problems is shown below.

Example	Brief Description
5.1	If $L_b = 0$ or if $L_b \leq L_p$, the <u>section modulus table</u>(LRFD page 3-13) is the best available beam design aid. The moment of inertia table(LRFD page 3-20) is helpful if deflection is the controlling design criterion. Uniform load on a simply supported beam.
5.2	If $C_b = 1$ and $L_b > L_p$, the <u>beam charts</u>(LRFD page 3-57 to 3-102) are the best available beam design aid. Uniform load on a simply supported beam.
5.3	If $C_b > 1$ and $L_b > L_p$, the section modulus table and the beam charts are used. Uniform load on a simply supported beam.
5.4	Indeterminate beam example.
5.5	LRFD A5.1(page 6-26) is used in Example 5.4 solution.
5.6	Simply supported beam with a concentrated load.
5.7	For the W section chosen in Example 5.6, design the bearing plates at the concentrated load and at the supports.

EXAMPLE 5.1_____

A simply supported W section beam is subjected to X-axis bending due to the following uniformly distributed service loads:
   (**dead load** -- includes estimated beam weight), **D** = 0.8 k/ft
   (**live load**), **L** = 1.4 k/ft
Beam **span** length = 30 feet. Assume the compression flange can be laterally braced such that $L_b \leq L_p$. The *limiting deflection due to service live load* is **Span/360** = (360 in.)/360 = 1.00 inches.

Select the lightest W section for:
   a) A36 steel
   b) A572 Grade 50 steel
   c) A572 Grade 65 steel
For each selected W section, show the <u>maximum</u> value of $L_b = L_p$.

SOLUTION--The factored loading is(see LRFD page 6-25):
$w_u = 1.2\,D + 1.6\,L = 1.2*0.8 + 1.6*1.4 = 3.2$ k/ft

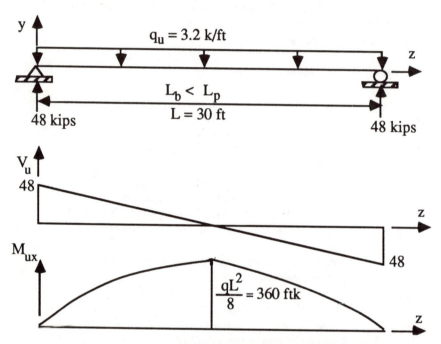

The design requirements are summarized in LRFD A5(page 6-26).
In this example problem, the applicable <u>design requirements</u> are:

1. $\phi_b M_{nx} \geq M_u$ in each region of $L_b \leq L_p$ [see LRFD F1(page 6-42)]
   and the **width-thickness ratios** *of the compression elements*
   <u>must not exceed</u> the applicable values of λ_p on LRFD page 6-32.

2. $\phi_v V_n \geq V_u$ [see LRFD F2(page 6-45)].

3. (**Maximum deflection** *due to <u>service live load</u>*) $\leq$ **1.00 in.**

<u>SOLUTION (a)</u> Select the lightest W section for $F_y = 36$ ksi steel

Assume that the *design bending strength* <u>controls the selection</u> of the lightest W section. Since $L_b \leq L_p$, if:

$$0.5 b_f/t_f \leq \left[\lambda_p = 65/\sqrt{F_y} = 65/\sqrt{36} = 10.8 \right]$$

$$h_c/t_w \leq \left[\lambda_p = 640/\sqrt{F_y} = 640/\sqrt{36} = 106.7 \right]$$

the selection criterion is: $\phi_b M_{px} \geq (M_{ux} = 360 \text{ ftk})$

See LRFD page 3-15(<u>students should read pages 3-9,10</u>):
Try W24X55 ($\phi_b M_{px} = 362$ ftk) $\geq$ ($M_u = 360$); $L_p = 5.6$ ft.

If a value of ϕM_{px} is given in the section modulus table for a W section, $b_f/(2t_f) \leq 65/\sqrt{F_y}$ and $h_c/t_w \leq 640/\sqrt{F_y}$. In this first example problem, the author will check these compression element requirements to illustrate the procedure which was used by the person who prepared the section modulus table to determine that a section's flange and web qualified for the listed ϕM_{px} value.

See LRFD page 1-24:
W24X55 $d = 23.57$; $t_w = 0.395$; $I_x = 1350$
$[0.5 b_f/t_f = 6.9] \leq [65/\sqrt{36} = 10.8]$ (flange OK for $\phi_b M_{px}$)
$[h_c/t_w = 54.6] \leq [640/\sqrt{36} = 106.7]$ (web OK for $\phi_b M_{px}$)

<u>Check shear</u>(see LRFD page 6-45):
$[h_c/t_w = 54.6] \leq [187\sqrt{5/36} = 69.7]$
$[\phi V_n = 0.9*0.6*36*23.57*0.395 = 181 \text{ kips}] > [V_u = 48]$ OK

<u>Check deflection</u>(see LRFD page 3-130 case 1):
Service live load, $w = 1.4$ k/ft; limiting deflection $= 1.00$ in.

$$\text{max.}\Delta = 5wL^4/(384EI)$$
$$= 5*(1.4\text{k/ft})*(30\text{ft.})^4*(12\text{in.ft})^3/[384*(29000\text{ksi})*(1350\text{in})]$$
$$= 0.652 \text{ in.}$$

(max.$\Delta = 0.652$ in.) $\leq$ (limiting deflection $= 1.00$ in.)   OK

Use W24X55 (it satisfies all design requirements) and $L_b \leq 5.6$ ft.

Note that <u>for this example problem,</u>
if $I_x < [0.652*1350 = 880$ in^4), max.$\Delta > 1.00$ in.;
that is, the deflection criterion is violated if $I_x < 880$ in^4.

<u>SOLUTION b)</u> Select lightest W section for $F_y = 50$ ksi steel
LRFD page 3-15:
Try W18X50 ($\phi_bM_{px}= 379$ ftk) $\geq$ ($M_{ux} = 360$); $L_p= 5.8$ ft.
LRFD page 3-20: [$I_x = 800$] < 880 NG
(No Good -- violates deflection criterion)

Try W21X50 ($\phi_bM_{px} = 413$ ftk) $\geq$ ($M_{ux} = 360$); $L_p = 4.6$ ft.
[$I_x = 984$] ≥ 880 OK(satisfies deflection criterion)
[$0.5b_f/t_f = 6.1$] $\leq [65/\sqrt{50} = 9.19$]   flange OK for ϕ_bM_{px}
[$h_c/t_w = 49.4$] $\leq [640/\sqrt{50} = 90.5$]   web OK for ϕ_bM_{px}

<u>Check shear</u>:
[$h_c/t_w = 49.4$] $\leq [187\sqrt{5/50} = 59.1$]
($\phi_vV_n = 0.9*0.6*50*20.83*0.380 = 214$ kips) $\geq$ ($V_u = 48$)   OK

Use W21X50 (it satisfies all design requirements) and $L_b \leq 4.6$ ft.

<u>SOLUTION (c)</u> Select lightest W section for $F_Y = 65$ ksi steel
LRFD page 3-16 does not list ϕM_{px} for 65 ksi steel, so we must do as
follows:
need: [$\phi_bM_{px} = 0.9*Z_x*(65$ ksi)] $\geq$ [$M_{ux} = 360$ ftk $= 4320$ ink]
      $Z_x \geq [4320/(0.9*65) = 73.8$ in] for design bending strength
      $I_x \geq 880$ in. for deflection criterion
LRFD page 3-16:
Try W18X40 ($Z_x = 78.4$) ≥ 73.8, but ($I_x = 612$) < 880 NG

LRFD page 3-20:
Try W21X50 ($I_x = 984$) ≥ 880, and ($Z_x = 110$) ≥ 73.8 OK

<u>Check flange and web:</u>
[$0.5b_f/t_f = 6.1$] $\leq [65/\sqrt{65} = 8.06$]   OK
[$h_c/t_w = 49.4$] $\leq [640/\sqrt{65} = 79.4$]   OK

See LRFD page 6-43 for definition of L_p:

$$\left[L_p = \frac{300 r_y}{\sqrt{F_y}}\right] = \frac{300 * 1.30}{\sqrt{65}} = 48.37 \text{ in.} = 4.03 \text{ ft.}$$

Check shear:

$[h_c/t_w = 49.4] \leq [187\sqrt{5/65} = 51.9]$

$(\phi_v V_n = 0.9*0.6*65*20.83*0.380 = 278 \text{ kips}) \leq (V_u = 48)$ OK

Use W21X50 and $L_b \leq 4.03$ ft.

Summary comments for EXAMPLE 5.1:

For 36 ksi, *design bending strength* controlled the selection of W24X55, but for 50 ksi and 65 ksi steel.*the limiting deflection criterion* controlled the selection (W21X50 was needed for 65 ksi and 50 ksi and there is no advantage in using 65 ksi steel in this problem). The selected section for 50 ksi is only 5 lb/ft lighter than the section chosen for 36 ksi steel. The actual cost of the section chosen for 36 ksi may be cheaper than the section chosen for 50 ksi.

Note that L_p decreased(L_p = 5.6, 4.6, 4.0 ft) as the steel grade increased(36, 50, 65 ksi). Therefore, the cost of the lateral braces increases as the steel grade increases.

LRFD page 3-20 is helpful when deflection controls the selection of the section.

Note that shear did not come close to controlling any W section selection in this example. For the W24X55 F_y = 36 ksi case, $V_u /(\phi V_n)$ = (48 kips)/(181 kips) = 0.265; that is, only 26.5% of the available shear design strength was used in satisfying the design requirement for shear. For a simply supported beam subjected to only a uniformly distributed load and if the W section design bending strength is ϕM_{px}, the condition at which shear and bending simultaneously control the section selection can be determined as follows:

$[\text{max. } V_u = w_u L/2] = \phi_v V_n$

$\quad w_u L = 2 * \phi_v V_n$

$[\text{max. } M_{ux} = w_u L^2/8] = \phi_b M_{px}$

$\quad w_u L * (L/8) = \phi_b M_{px}$

$L = 8 * \phi_b M_{px}/(2 * \phi_v V_n) = 4 * \phi_b M_{px}/\phi_v V_n$

is the value of L at which shear and bending simultaneously control the section selection if design strength controls. For the W24X55 F_y = 36

175

ksi case, for example, L = 4*(362 ftk)/(181 kips) = 8.00 ft. and the value of w_u which makes shear and bending simultaneously control the section selection can be obtained from:

$$[w_u L/2 = w_u*(8 \text{ ft})/2] = [\phi_v V_n = 181 \text{ kips}]; \quad w_u = 45.25 \text{ k/ft}.$$

If L < 8.00 ft, *shear* <u>controls the selection</u> of a W24X55 F_y = 36ksi for a uniformly loaded(w_u = 45.25 k/ft), simply supported beam. If L > 8.00 ft, *moment* <u>controls the section selection</u> and w_u < 45.25 k/ft. **Conclusion:** *shear* does not control the section selection for a simply supported beam subjected only to a uniformly distributed load unless the beam is very short and very heavily loaded. Hereafter, unless the span of a simply supported and uniformly loaded beam is less than 10 ft., the author will not show the design check for shear of a W section.

EXAMPLE 5.2 _____

Repeat EXAMPLE 5.1 assuming that lateral bracing can only be provided at the supports and at the L/3 points; L_b = (30 ft)/3 = 10 ft.

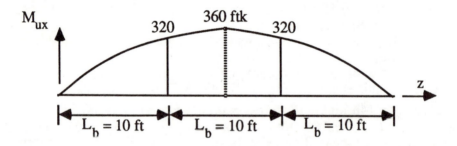

SOLUTION--See LRFD page 6-43: For the middle L_b = 10 ft region on the moment diagram, C_b = 1 since the maximum moment(360 ftk) exceeds the moment(320 ftk) at the ends of the unbraced length. For the end L_b = 10 ft regions, C_b = 1.75 and M_{ux} = 320. Obviously, we must choose the W section for the L_b = 10 ft. region where C_b = 1 and M_{ux} = 360 ftk.

In the summary comments for EXAMPLE 5.1, it was noted that the maximum value of L_p was 5.6 ft. for the W24X55 F_y = 36 ksi case. Since L_b = 10 ft. for EXAMPLE 5.2, it is likely that $L_b > L_p$. See Figure 5.9(page **5-8.1**). In the $L_p < L_b < L_r$ region, the *nominal bending strength* <u>varies linearly</u> and the information given on LRFD page 3-15 could be used to select the lightest W section, but a selection by inspection would not be possible. Fortunately, when C_b = 1, the beam charts on LRFD pages 3-53 to 3-55 can be used to select the lightest W section by inspection. For each W section in these beam charts, the shape of the plotted information is similar to Figure 5.9 (page **5-8.1**), but the information continues across several LRFD pages. Students should read LRFD pages 3-53 to 3-55 before using the described beam charts. It should be noted that each section listed in these beam charts satisfies the applicable limiting λ_p width-thickness ratios for compression elements shown on LRFD page 6-32. Therefore, the web and compression flange of each section qualify everywhere on the applicable curve in the beam charts and the author will not show any flange or web check calculations when these beam charts are used to select a section.

SOLUTION a) Select the lightest W section for 36 ksi steel
See LRFD page 3-68:
Locate the point whose coordinates are: $(L_b = 10$ ft, $M_{ux} = 360$ ftk). Any curve to the right of and above this point is a satisfactory solution since $\phi M_{px} \geq (M_{ux} = 360$ ftk). Solid lined curves indicate the lightest choice in each region. Therefore, the solid line closest to but above and to the right of the plotted point is the lightest section which satisfies the design requirement for bending strength.
Try W21X62 ($\phi_b M_{px} = 363$ ftk at $L_b = 10$ ft) $\geq (M_{ux} = 360$ ftk)
$(I_X = 1330) \geq 880$ (deflection criterion is satisfied)
Use W21X62

SOLUTION b) Select the lightest W section for 50 ksi steel.
See LRFD page 3-92:
Try W18X55 ($\phi_b M_{px} = 370$ ftk at $L_b = 10$ ft) $\geq (M_{ux} = 360)$
$(I_X = 890) \geq 880$ (deflection criterion is satisfied)
Use W18X55

SOLUTION c) Select lightest W section for 65 ksi steel.
There are not any $C_b = 1$ curves available for 65 ksi steel.
Use $C_b = 1$ curves for $F_y = 50$ ksi and $M_{ux} = (50/65)*360 = 277$ ftk to choose a trial section at $L_b = 10$ ft for 65 ksi steel.
Try W21X44 $F_y = 65$ ksi
Check deflection:
See LRFD page 3-94: W21X44 $(I_X = 843) < 880$ NG
Try W21X50 $(I_X = 890) \geq 880$ (deflection criterion is satisfied)

We must compute $\phi_b M_{nx}$ to determine if $\phi_b M_{nx} \geq (M_{ux} = 360$ ftk).

See LRFD page 6-43:

$$\left[L_p = \frac{300 r_y}{\sqrt{F_y}}\right] = 300*1.35/\sqrt{65} = 50.2 \text{ in.} = 4.19 \text{ ft.}$$

See LRFD page 1-136 for J and C_w values of W21X50, LRFD page 1-333 for G = 11,200 ksi, and LRFD page 1-24 for the other needed properties of W21X50.

Return to LRFD page 6-43:

Eqn(F1-8): $X_1 = (\pi/94.5)*\sqrt{29000*11200*1.14*14.7/2} = 1734$

Eqn(F1-9): $X_2 = (4*2570/24.9)*[94.5/(11200*1.14)^2] = 0.022616$

Eqn(F1-6): $L_r = [1.30*1734/(65-10)]*\sqrt{1 + \sqrt{1+0.022616*(65-10)}}$
$\qquad\qquad = 64.8 \text{ in.} = 5.40 \text{ ft.}$

Eqn(F1-7): $M_{rx} = (65 - 10)*94.5 = 5198 \text{ ink} = 433 \text{ ftk}$

(L_b = 10 ft) > (L_r = 5.40 ft),
see LRFD page 6-44[Eqns(F1-12,F1-13)]

Eqn(F1-13): $M_{cr} = 1.0*[\pi/(10*12)]*\sqrt{a + b}$

$\qquad\qquad a = 29000*24.9*11200*1.14 = 9.22 \times 10^9$

$\qquad\qquad b = [29000\pi/(10*20)]^2*24.9*2570 = 3.69 \times 10^{10}$

$\qquad\qquad \sqrt{a + b} = 214,723$

$\qquad\qquad (M_{cr} = 5621 \text{ ink} = 468 \text{ ftk}) > (M_{rx} = 1.0*433 = 433)$

$(\phi_b M_{nx} = 0.9*433 = 390 \text{ ftk}) \geq (M_{ux} = 360)$ $\qquad\qquad$ OK

Use W21X50

EXAMPLE 5.3_____

Repeat EXAMPLE 5.1 assuming that lateral braces can only be provided at the supports and at midspan; $L_b = 15$ ft.

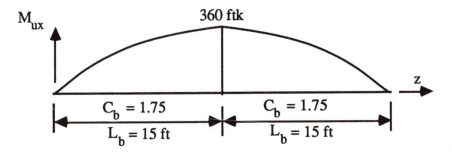

SOLUTION--See Figure 5.10(page **5-8.2**).
For $C_b > 1$ and $L_b > L_p$, the *X-axis nominal bending strength* is:

$$[M_{nx} = C_b*(M_{nx} \text{ for } C_b = 1)] \leq \phi M_{px}$$

Therefore, when $C_b > 1$, the author recommends the selection procedure given in the following example.

<u>SOLUTION a)</u> Select lightest W section for 36 ksi steel

The author's recommended selection procedure is:

1. Begin with the section modulus table since $\phi_b M_{px}$ will be needed in the selection from the beam charts. Do not worry about L_b and select the lightest section in the section modulus table that satisfies the design requirement for bending strength and record the value of $\phi_b M_{px}$ for the selected section.

   LRFD page 3-15, W24X55 ($\phi_b M_{px} = 362$ ftk) ≥ ($M_{ux} = 360$ ftk)
   In our example, we must check deflection:
   W24X55 ($I_x = 1350$) ≥ 880   OK

2. On LRFD page 3-70, locate the point whose coordinates are:
   [$L_b = 15$,($M_{ux} /C_b = 360/1.75 = 206$)]; look to the right of and above this point for the W24X55 curve. Since the W24X55 curve is found in the described search region, W24X55 $F_y = 36$ ksi with $L_b = 15$ and $C_b = 1.75$ is the lightest possible choice that satisfies the design requirement for bending strength.

179

NOTE: If the W24X55 curve could not be found in the described search region, *lateral-torsional buckling* would control the selection and we would choose the lightest section in the described search region(in our example, the chosen section would have to be checked for deflection.)

Use W24X55.

Although it is not necessary to do so, the author will verify that the selected W24X55 section indeed satisfies the design requirement for bending strength when $C_b = 1.75$ and $L_b = 15$ ft.

LRFD page 3-15: W24X55 $F_y = 36$ ksi

$L_r = 16.6$ ft; $L_p = 5.6$ ft; $\phi_b M_{rx} = 222$ ftk; $\phi_b M_{px} = 362$ ftk

$(L_p = 5.6) < (L_b = 15) < (L_r = 16.6)$

On LRFD page 6-43 we find that $\phi_b M_{nx}$ is the smaller of $\phi_b M_{px} = 362$ ftk and $C_b*[Eqn(F1-3)]$ where $\phi_b = 0.9$

Note that 0.9 is already contained in $\phi_b M_{rx} = 222$ and $\phi_b M_{px} = 362$.
$1.75*[362 - (362 - 222)*(15 - 5.6)/(16.6 - 5.6)] = 424$ ftk

$\phi_b M_{nx}$ is the smaller of 424 ftk and 362 ftk.

Therefore, $[\phi_b M_{nx} = 362 \text{ ftk}] \geq [M_{ux} = 360]$ and this verifies that the section selected by the author's recommended design procedure is okay.

SOLUTION b) Select the lightest W section for 50 ksi steel

LRFD page 3-15: W18X50 $(\phi_b M_{px} = 379 \text{ ftk}) \geq (M_{ux} = 360)$
However, $(I_X = 800) < 880$ NG

W21X50 $(I_X = 984) \leq 880$; $(\phi_b M_{px} = 413 \text{ ftk}) \geq (M_{ux} = 360)$

LRFD page 3-94:
locate point $[L_b = 15, (M_{ux}/C_b = 360/1.75 = 206)]$
The W21X50 curve is barely to the right of and above this point.
Therefore, use W21X50.

<u>SOLUTION c)</u> Select the lightest W section for 65 ksi steel
(50/65)*360 = 277 ftk

LRFD page 3-16: W18X40 ($\phi_b M_{px}$ = 294 ftk) ≥ 277
$(I_X = 612) < 880$
NG for deflection criterion.

W21X50 $(I_X = 984) ≥ 880$

$(\phi_b M_{px} = 413) ≥ 277$

Try W21X50
On LRFD page 3-96, W21X50 appears to right of L_b = 15 ft and above
(277 ftk.)/1.75 = 158 ftk which is a good sign, but since $(F_y = 65) ≠ 50$
<u>we must perform</u> the *bending requirement check* as follows:

LRFD Eqn(F1-13):
$$M_{cr} = \frac{1.75\pi}{180}\sqrt{29000*24.9*11200*1.14+(29000\pi/180)^2*24.9*2570}$$
M_{cr} = 4888 ink = 407 ftk
$M_{rx} = (F_y - 10\ ksi)*S_X = (65 - 10)*94.5 = 5197.5$ ink = 433 ftk
$(M_{cr} = 407\ ftk) ≤ (C_b M_{rx} = 1.75*433 = 758)$
$(\phi_b M_{nx} = 0.9*407 = 366\ ftk) ≥ (M_{ux} = 360\ ftk)$

Use W21X50

EXAMPLE 5.4_____

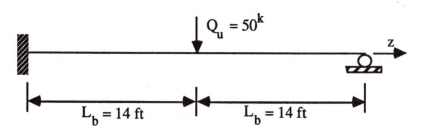

Lateral braces are provided only at the supports and at midspan. Ignore
beam weight. There is not any limiting deflection criterion to be
satisfied. Find lightest W section of A36 steel .

SOLUTION--See LRFD page 3-134 case 13

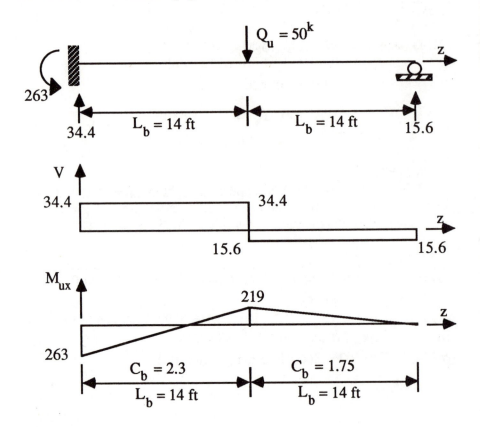

For the first $L_b = 14$ ft. region on the moment diagram, C_b is the smaller of 2.3 and
$$[1.75 + 1.05*(219/263) + 0.3*(219/263)^2 = 2.83]$$
$C_b = 2.3$

For the second $L_b = 14$ ft. region on the moment diagram, $C_b = 1.75$

LRFD page 3-15: W18X50 ($\phi_b M_{px} = 273$ ftk) $\geq (M_{ux} = 263)$

W21X50 ($\phi_b M_{px} = 297$ ftk) $\geq (M_{ux} = 263)$

Enter LRFD page 3-74 at $L_b = 14$ ft. with the larger of:
    (263 ftk)/2.3 = 114 ftk   (for first L_b region)
    (219 ftk)/1.75 = 125 ftk   (for second L_b region)

182

W18X50 and W21X50 appear to the right of and above the point (L_b = 14 ft, 125 ftk).

Although it is not necessary to do so, verify W18X50.
On LRFD page 3-70 at L_b = 14 ft., [$\phi_b M_{nx}$ = 221 ftk for C_b = 1].
For C_b = 1.75, $\phi_b M_{nx}$ is the smaller of:

$$\phi_b M_{px} = 273 \text{ ftk}$$
$$C_b*[\phi_b M_{nx} \text{ for } C_b = 1] = 1.75*221 = 387 \text{ ftk.}$$

($\phi_b M_{nx}$ = 273 ftk) ≥ (M_{ux} = 219)   OK for second L_b region

For C_b = 2.3, $\phi_b M_{nx}$ is the smaller of:

$$\phi_b M_{px} = 273 \text{ ftk}$$
$$C_b*[\phi_b M_{nx} \text{ for } C_b = 1] = 2.3*221 = 508 \text{ ftk.}$$

($\phi_b M_{nx}$ = 273 ftk) ≥ (M_{ux} = 263)   OK for first L_b region

Use either W18X50 or W21X50

EXAMPLE 5.5 _____

Repeat EXAMPLE 5.4 accounting for LRFD A5.1(page 6-26).

SOLUTION--See LRFD A5.1(page 6-26). The maximum possible reduction in the moment at the left support is 0.1*263 = 26.3 ftk. Note that we can not change the moment at the right support in this example. Try reducing 263 ftk by 26 ftk and increasing 219 ftk by (26+0)/2 = 13 ftk.
263 - 26 = 237 ftk and 219 + 13 = 232 ftk.
The adjusted moment diagram is:

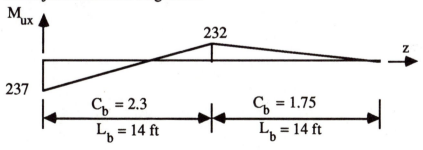

LRFD page 3-15: W21X44 ($\phi_b M_{px}$ = 258 ftk) $\geq$ (M_{ux} = 237)

237/2.3 = 103 and 232/1.75 = 132.6
Enter LRFD page 3-74 at (L_b = 14, M_{ux}/C_b = 132.6)
W21X44 is to the right of and above this point; W21X44 is OK.

Use W21X44.

For EXAMPLE 5.4, either a W18X50 or W21X50 was needed. Thus, by accounting for LRFD A5.1, we can use W21X44 and the savings is 6 lb/ft which is a weight saving of (6/50)*100 = 12%.

EXAMPLE   5.6_____

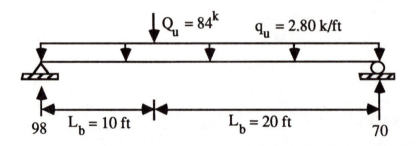

Lateral braces are provided only at the supports and at the concentrated load. There is not any limitation on deflection. The estimated factored beam weight is included in w_u = 2.80 k/ft . Find the lightest W section of A36 steel that satisfies the LRFD Specifications for bending strength.

184

SOLUTION

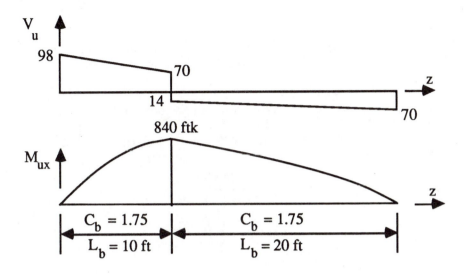

LRFD page 3-14: W30X99 ($\phi_b M_{px}$ = 842 ftk) ≥ (M_{ux} = 840)

Enter LRFD page 3-66 at L_b = 20 ft. with M_{ux}/C_b = 840/1.75 = 480
W30X99 lies to right of and above this point.
W30X99 is OK for design bending requirement .

Use W30X99.

EXAMPLE   5.7_____

In EXAMPLE 5.6, suppose the P_u = 84 kips is due to the reaction of a
W24X84 sitting on the top surface of the W30X99 chosen in
EXAMPLE 5.6 and the W30X99 is supported at each end on an 8 inch
thick concrete wall.
Concrete grade: (f_c' = 3 ksi)
Steel grade: F_y = 36 ksi

1. Is a bearing plate needed between the W24X84 and W30X99?
    If the answer is yes, specify the value of bearing length, N, for
    each W section.
2. Is a bearing plate needed between the W30X99 and the concrete
    walls? If the answer is yes, design the bearing plates.

185

SOLUTION--1a. Check web of W30X99 beneath W24X84 without a
bearing plate.

LRFD page 1-24: W24X84
$k_1 = 15/16$; $k = 25/16$; $t_w = 0.470$; $d = 24.10$; $t_f = 0.770$
$N = 2k_1 = 2*15/16 = 1.875$ for W24X84 on web of W30X99

LRFD page 1-20: W30X99
$k_1 = 1$; $k = 23/16$; $t_w = 0.520$; $d = 29.65$; $t_f = 0.670$
$d_c = h = T = 26.75$; $b_f = 10.45$
$N = 2k_1 = 2*23/16 = 2.6875$ for W30X99 on web of W24X84

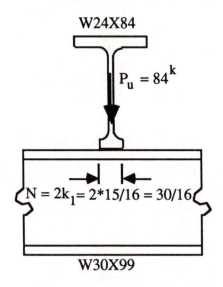

LRFD K1.3(page 6-77) *Local web yielding* of W30X99
required: [$\phi R_n = 1.0*$[LRFD Eqn(K1-2)]] $\geq$ 84 kips
[$\phi R_n = 1.0*(5*23/16 + 30/16)*36*0.520 = 170$ kips] $\geq$ 84   OK

LRFD K1.4(page 6-78) *Web crippling* of W30X99
required: [$\phi R_n = 0.75*$[Eqn(K1.4)]] $\geq$ 84 kips
$$\phi R_n = 0.75*135*(0.520)^2*\left[1+3*\frac{1.875}{29.65}*\left(\frac{0.520}{0.670}\right)^{1.5}\right]\sqrt{\frac{36*0.670}{0.520}}$$
[$\phi R_n = 210$ kips] $\geq$ 84   OK

LRFD K1.5(page 6-78) *Sidesway web buckling* of W30X99
This specification is applicable if the bottom flange of W30X99 is not
laterally braced at the $P_u = 84$ kip location. The author shows the
calculations for illustration purposes.

Maximum web flexural stress $= \dfrac{M_X\, y}{I_X}$

$$= 840*12*(29.65/2 - 23/16)/3990$$
$$= 33.8 \text{ ksi} < (F_y = 36 \text{ ksi})$$

required: $[\phi R_n = 0.85*[2*[\text{LRFD Eqn(K1-6)}] \geq 84$ kips

$\phi R_n = 0.85*2*12000*[(0.520)^3/26.75]*[1+0.4*(51.9*10.45/360)^3]$

$[\phi R_n = 254 \text{ kips}] \geq 84$ OK

A bearing plate is not needed to satisfy any of the applicable LRFD
Specifications for the web of the W30X99 at the $P_u = 84$ kip location.

SOLUTION 1b. Check the W24X84 web without a bearing plate.
Assume the 84 kip reaction is an end reaction.

W24X84

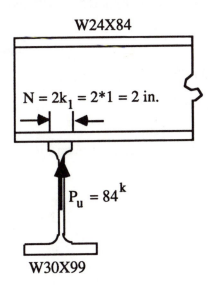

$N = 2k_1 = 2*1 = 2$ in.

$P_u = 84^k$

W30X99

LRFD K1.3(page 6-77) *Local web yielding* of W24X84
required: $[\phi R_n = 1.0*[\text{LRFD Eqn(K1-3)}]] \geq 84$ kips
$[\phi R_n = 1.0*(2.5*25/16 + 2)*36*0.470 = 99.9] \geq 84$ OK

LRFD K1.4(page 6-78) *Web crippling* of W24X84

required: $[\phi R_n = 0.75*[\text{LRFD Eqn(K1-5)}]] \geq 84$ kips

$$\phi R_n = 0.75*68*(0.470)^2*\left[1+3*\frac{2}{24.10}*\left(\frac{0.470}{0.770}\right)^{1.5}\right]*\sqrt{\frac{36*0.770}{0.470}}$$

$[\phi R_n = 96.8$ kips$] \geq 84$ OK

LRFD K1.5 *Sidesway web buckling* is not applicable since the top flange of the W24X84 must be laterally braced at the member end where the W24X84 is being checked due to the reaction.

A bearing plate is not needed to satisfy any of the applicable LRFD Specifications for the W24X84 web requirements.

SOLUTION 2. Is a bearing plate needed at the left beam reaction?

a) without a bearing plate
Assume that the bearing length of the W30X99 web on the concrete wall is 6 in. and <u>assume</u> the *bearing width* = $2K_1$ = 2*1 = 2 in. See author's NOTE at the end of the solution for the bearing width assumption.

LRFD J9(page 6-75) $A_2 = A_1$

required: $[\phi_c P_p = 0.6*(0.85f_c' A_1)] \geq 98$ kips

$[\phi_c P_p = 0.6*0.85*3*6*2 = 18.4] < 98$

A bearing plate must be designed to protect the concrete wall. Use the recommended design procedure on LRFD page 3-49.

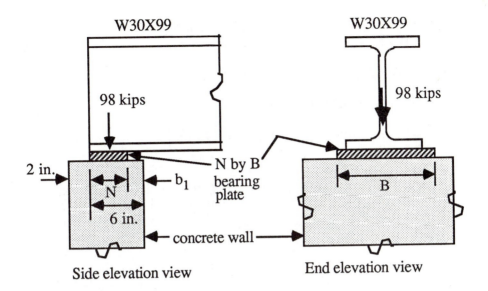

W30X99

W30X99

98 kips

98 kips

2 in.

N by B
bearing
plate

b_1

N

6 in.

B

concrete wall

Side elevation view

End elevation view

See LRFD J3.10(page 6-72)
Use $N \geq (2*1.5 = 3$ in.) to accommodate up to 1.125 in. diameter
anchor bolt.
$N = $ [(bearing length of W30X99 web on wall) - b_1]
$b_1 \geq 1$ in. (author's choice); Actual b_1 choice is made later.
$N \leq [6 - 1 = 5$ in.]

See the first two formulas on LRFD page 3-50.
See LRFD page 3-30 for the W30X99 properties needed below.

Local web yielding: $N \geq (R - \phi R_1)/(\phi R_2)$
$$N \geq [(98 - 67.3)/18.7 = 1.64 \text{ in.}]$$

Web crippling: $N \geq (R - \phi_r R_3)/(\phi_r R_4)$
$$N \geq [(98 - 93.9)/6.50 = 0.631 \text{ in.}]$$
Use $B \geq (b_f = 10.45$ for W30X99) and $6 \geq N \geq 3$
Try $B = 11$ in.; $A_1 = B*N = 11*N$

LRFD page 6-75:
required: $\sqrt{A_2/A_1} \leq 2$; therefore, $A_2 \leq 4 A_1$ is required.
Try $A_2 = 4*A_1 = 4*(11*N) = 44*N$
LRFD page 3-50: $A_1*A_2 \geq [R/(\phi_c*0.85f_c')]^2$ is required

$$(11*N)*(44*N) \geq [98/(0.6*0.85*3)]^2$$
$$N \geq 2.91 \text{ in.}$$

189

LRFD page 3-50 states that preferably B and N should be in full inches. For N = 3 in. and B = 11 in.; $A_1 = 3*11 = 33$ in^2

LRFD page 3-50: n = (B/2 - k) = 11/2 - 23/16 = 4.0625 in.

Use t $\geq \left[n\sqrt{2.22R/(A_1F_y)} \right]$

   t $\geq \left[4.0625\sqrt{2.22*98/(33*36)} = 1.74 \text{ in.} \right]$

See LRFD page 1-121:
(Width = 3 in.) $\leq$ 6 and (thickness = 1.75 in.) $\geq$ 0.203; the bearing plate is classified as a bar. Preferred thickness increment = 1/8 inch. N X B X t = 3 X 11 X 1.75 bearing plate at each reaction is acceptable to protect the concrete wall. Bearing plate volume = 57.8 in^3

Alternatively, we could choose N = 4 ; B = 11; $A_1 = 4*11 = 44$ in^2

Then t $\geq \left[4.0625\sqrt{2.22*98/(44*36)} = 1.506 \text{ in.} \right]$

and a 4 X 11 X 1.5 bearing plate would be acceptable. Volume= 66 in^3

Use 3 X 11 X 1.75 bearing plate

NOTE: Without a bearing plate, the author chose to say that the *bearing width* = $2k_1$ = 2 in. The reason is that the W30X99 flange is not perfectly flat; the flange either curls up or down at each cross section. If the flange curls down, it will flatten out when the beam reaction occurs. However, if the flange curls up, the nut on the anchor bolts would have to be tightened enough to make the flange flatten out. For a simply supported beam, this is not desirable on one end of the beam since friction between the concrete wall and the bottom flange of the W30X99 would prevent the assumed roller end from sliding until friction was overcome. Thus, when the bottom flange curls up, only the $2k_1$ width portion of the W30X99 beam is in contact with the wall along the 6 inch bearing length. If we deemed it appropriate to assume that the bearing width, B = (b_f = 10.5 in.), then:

($\phi_cP_p = 0.6*0.85*3*6*10.5 = 96.6$ kips) $\approx$ 98   OK
n = B/2 - k = 10.5/2 - 23/16 = 3.8125 in.

need: $t_f \geq [3.8125\sqrt{2.22*98/(6*10.5*36)} = 1.18$ in.]
However, (t_f = 0.670 in.) < 1.18 in. and the W30X99 flange thickness is not thick enough to serve as the bearing plate.

5.11 BIAXIAL BENDING OF SYMMETRIC SECTIONS

The design requirement for biaxial bending of a W section beam is LRFD Eqn(H1-1b) with $P_u = 0$ and $\phi_b = 0.9$; see LRFD page 6-47.

that is, $\dfrac{M_{ux}}{\phi_b M_{nx}} + \dfrac{M_{uy}}{\phi_b M_{ny}} \leq 1.00$ is required.

EXAMPLE 5.8_____

In Figure 5.18, a W36X170 $F_y = 50$ ksi is used as a simply supported beam with regard to both principal axis. Lateral braces are provided only at the supports. The beam weight is negligible and the only load is 100 kips concentrated at midspan. The load passes through the shear center of the cross section, but the load is not parallel to either principal axis of the cross section. As shown in Figure 5.18, the components of the 100 kip load cause bending moments to occur about both principal axes. Does the described beam satisfy the design requirement for biaxial bending?

SOLUTION
$M_{ux} = 739$ ftk

$C_b = 1$; $L_b = 30$ ft; $\phi_b M_{nx} = 1236$ ftk [from LRFD page 3-85]
$M_{uy} = 130$ ftk
For *Y-axis design bending strength*, see text page **5-9**(Eqn 5.7) and LRFD F1.7(page 6-45):

$\phi_b M_{ny} = 0.9 Z_Y F_y = 0.9*(83.8 \text{ in}^3)*(50 \text{ ksi}) = 3771$ ink $= 314$ ftk

$\left[\dfrac{739}{1236} + \dfrac{130}{314} = 0.598 + 0.414 = 1.012 \right] > 1.00 \qquad$ NG

Some structural designers might say $1.012 \approx 1.00 \qquad$ OK

However, in the process of finding $\phi_b M_{nx}$, the author noticed that a W27X146 $F_y = 50$ ksi has slightly more X-axis design bending strength than the W36X170. Maybe the W27X146 will satisfy the design requirement for biaxial bending.
Try W27X146 $F_y = 50$ ksi

$C_b = 1$; $L_b = 30$ ft; $\phi_b M_{nx} = 1250$ ftk [from LRFD page 3-85]

$\phi_b M_{ny} = 0.9*97.5*50/12 = 366$ ftk

$\left[\dfrac{739}{1250} + \dfrac{130}{366} = 0.591 + 0.355 = 0.946 \right] \leq 1.00 \quad$ OK, use W27X146

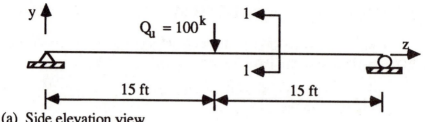

(a) Side elevation view

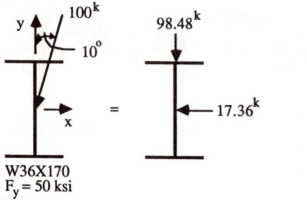

W36X170
$F_y = 50$ ksi

(b) Section 1-1

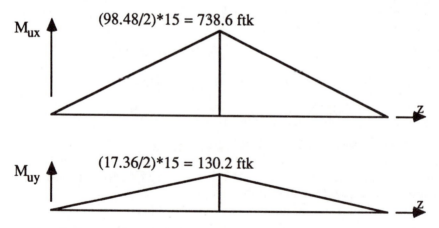

(c) Factored moment diagrams for principal axes

Figure 5.18 W section subjected to biaxial bending

5.12 BENDING OF UNSYMMETRIC SECTIONS

A single angle with unequal legs and a Z section are examples of an unsymmetric section. See LRFD F1.7(page 6-45) which states that LRFD Appendix F1.7(page 6-92) applies for other cross section types. However, LRFD Table A-F1.1(pages 6-94 to 6-97) does not give any nominal strength parameters for any unsymmetric sections nor for an equal legged angle. Therefore, the LRFD Specifications do not provide the nominal bending strength definition for an unsymmetric section or for an equal legged angle used as a beam. The following example illustrates the author's opinion of an acceptable approach for an unsymmetric section used as a beam.

EXAMPLE 5.9_____

An L6X4X1/2 $F_y = 36$ ksi section is used as a simply supported beam with lateral supports provided only at the beam ends. The factored loading is 0.7 k/ft and passes through the shear center as shown below.

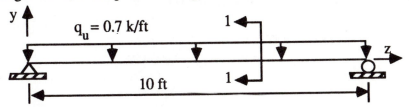

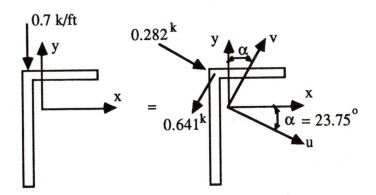

Section 1-1

Note that $q_u = 0.7$ k/ft is not parallel to either principal axis. The *nominal bending strength for each principal axis* must be

193

determined. Therefore, we must decompose the load into components parallel to each principal axis in order to find the **required bending moment for each principal axis.**

See text Appendix B which is used in the calculations shown below to locate the principal axes and to find the section properties for each principal axis of the L6X4X1/2.

See LRFD page 1-52 for the properties of L6X4X1/2:
$A = 4.75$; $I_X = 17.4$; $I_Y = 6.27$; $r_v = [r_{min} = r_z = 0.870]$

$\tan \alpha = 0.440$; $\alpha = 23.75$ degrees; $2\alpha = 47.5$ degrees

$I_v = [I_{min} = Ar_v^2] = 4.75*(0.870)^2 = 3.60$

$C = (I_X + I_Y)/2 = (17.4 + 6.27)/2 = 11.8$

$R = C - I_v = 11.8 - 3.60 = 8.2$

$R*\sin 2\alpha = 8.2*0.7372773 = 6.05$

$I_u = [I_{max} = C + R = 11.8 + 8.2 = 20.0]$

Product of Inertias

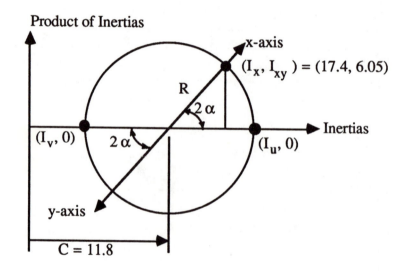

The components of the factored load are:

parallel to the u-axis: $(0.7 \text{ k/ft})*\sin\alpha = 0.282$ k/ft

parallel to the v-axis: $(0.7 \text{ k/ft})*\cos\alpha = 0.641$ k/ft

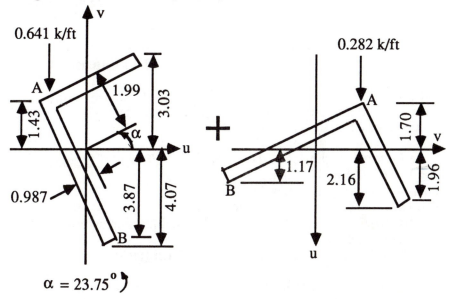

$\alpha = 23.75°$

In the above sketches, for example,

the distance from the u-axis to point B is:

$(6 - 1.99)*\cos\alpha - 0.5*\sin\alpha = 3.87$ in.

the distance from the v-axis to point B is:

$(6 - 1.99)*\sin\alpha - (0.987 - 0.5)*\cos\alpha = 1.17$ in.

The maximum factored bending moment for each principal axis is:

$M_u = (0.641 \text{ k/ft})*(10 \text{ ft}) /8 = 8.01$ ftk = 96.1 ink

$M_v = (0.292 \text{ k/ft})*(10 \text{ ft}) /8 = 3.53$ ftk = 42.4 ink

If the section remains elastic, the flexural formula is valid for each principal axis. Also, at any point on the cross section, the stress due to biaxial bending is the algebraic sum of the two stresses obtained by using the flexural formula. The *maximum compressive stress* due to biaxial bending <u>occurs at point A</u>:

$$\frac{M_u*1.43}{I_u} + \frac{M_v*1.70}{I_v} = \frac{96.1*1.43}{20.0} + \frac{42.4*1.70}{3.60}$$
$$= 6.87 + 20.02 = 26.9 \text{ ksi (compression)}$$

and since the residual stress at point A is 10 ksi (tension),

the combined stress is 26.9 - 10 = 16.9 ksi (compression)
which is less than the yield stress(36 ksi).

The *maximum tensile stress* due to biaxial bending occurs at point B:

$$\frac{M_u*3.87}{I_u} + \frac{M_v*1.17}{I_v} = \frac{96.1*3.87}{20.0} + \frac{42.4*1.17}{3.60}$$
$$= 18.6 + 13.8 = 32.4 \text{ ksi (tension)}$$

and since the residual stress at point B is 10 ksi (compression),
the combined stress is 32.4 - 10 = 22.4 ksi (tension)
which is less than the yield stress(36 ksi).

If the factored loading is 1.2D + 1.6L = 0.7 k/ft and if **D = L**, then
D = L = 0.7/2.8 = 0.25 k/ft
Total service load, D + L = 2*0.25 = 0.5 k/ft.
The **combined stress at point A** is:
   [(0.5/0.7)*26.9 = 19.21] - 10 = 9.21 ksi (compression).
The **combined stress at point B** is:
   [(0.5/0.7)*32.4 = 23.14] - 10 = 13.14 ksi (tension).

The preceding elastic stress calculations were performed to aid in
the understanding of the following approach used by the author to
check the design bending strength.
LRFD Eqn(F1-13) with I_v substituted for I_y is valid for elastic lateral-
torsional buckling of an equal legged angle bending about the major
principal axis. The term due to warping torsion is negligible as shown
later. For the L6X4X1/2 angle, an equivalent equal legged angle is
conservatively the larger of the following b values:
   b = 2*3.03/√2 = 4.29 in. (say L4X4X1/2)
   b = (3.03 + 4.07)/√2 = 5.02 in. (say L5X5X1/2)

See LRFD page 6-32:
(b/t = 5/0.5 = 10) ≤ (λ_r = 76/√36 = 12.7)
If the L5X5X1/2 were used as a column, local buckling would not
occur before column buckling occurred. Therefore, local buckling
cannot occur before lateral-torsional buckling occurs for the L5X5X1/2
used as a beam subjected to bending about the major principal axis.

For an L4X4X1/2 bending about the major principal axis:
LRFD Eqn(F1-13) with C_b = 1, C_w = 0, J = 0.322 and

196

$$I_Y = [I_v = A*r_v^2 = 3.75*(0.782)^2 = 2.29 \text{ in}^4] \text{ gives:}$$

$$M_{cr} = \frac{\pi\sqrt{29000*2.29*11200*0.322}}{120} = 405.2 \text{ ink} = 33.8 \text{ ftk}$$

If we do not ignore $C_w = 0.366$(see LRFD page 1-144),
$M_{cr} = 405.6$ ink which is only 0.1% larger than 405.2 ink. Hence, the term due to warping torsion is negligible.

For an L5X5X1/2 bending about the major principal axis, LRFD Eqn(F1- 13) gives $M_{cr} = 54.4$ ftk and for L6X4X1/2 $M_{cr} = 48.2$ ftk. To be conservative, use $M_{cr} = 33.8$ ftk obtained for L4X4X1/2.
Since $M_{cr} = 33.8$ ftk is for elastic behavior, if $M_{cr} > M_{ru}$, we must use M_{ru} as the nominal bending strength for the major principal axis. M_{ru} is the smaller of:
$(I_u/3.30)*(F_y - 10 \text{ ksi})/(12 \text{ in/ft}) = (20.0/3.30)*(36 - 10)/12 = 14.3$ ftk
$(I_u/4.07)*(F_y + 10 \text{ ksi})/(12 \text{ in/ft}) = (20.0/4.07)*(36 + 10)/12= 18.8$ ftk
$\phi_b M_{nu} = 0.9*14.3 = 12.9$ ftk

For an L6X4X1/2 bending about the minor principal axis, lateral-torsional buckling cannot occur. The nominal bending strength is limited to M_{rv} which is the smaller of:
$(I_v./1.70)*(F_y + 10)/12 = (3.60/1.70)*46/12 = 8.13$ ftk
$(I_v /1.96)*(F_y + 10)/12 = (3.60/1.96)*46/12 = 7.05$ ftk
$\phi_b M_{nv} = 0.9*7.05 = 6.35$ ftk

Check the design requirement for biaxial bending:
$$\left[\frac{8.01}{12.9} + \frac{3.53}{6.35} = 0.621 + 0.556 = 1.177\right] > 1.00 \qquad \text{NG}$$

PROBLEMS

5.1 Simply supported beam. L = 36 ft; $q_u = 3$ k/ft.
No restriction on deflection.
Find the lightest acceptable W section for:
a) $F_y = 36$ ksi
b) $F_y = 50$ ksi
c) $F_y = 65$ ksi
d) $F_y = 100$ ksi
For each chosen W section, find L_p.

5.2 Solve problem 5.1 with the restriction that the deflection due to service live load = 1.25 k/ft is not to exceed L/240.

5.3 Solve problem 5.1 for $L_b = 12$ ft.

5.4 Solve problem 5.1 for $L_b = 18$ ft.

5.5 Simply supported beam. $L = 48$ ft; $q_u = 1.5$ k/ft.; $L_b \le L_p$.
No restriction on deflection.
Find the lightest acceptable W section for:
a) $F_y = 36$ ksi
b) $F_y = 50$ ksi
c) $F_y = 65$ ksi
d) $F_y = 100$ ksi
For each chosen W section, find L_p.

5.6 Solve problem 5.5 with the restriction that the deflection due to service live load = 0.625 k/ft is not to exceed L/240.

5.7 Solve problem 5.5 for $L_b = 16$ ft.

5.8 Solve problem 5.5 for $L_b = 12$ ft.

5.9 Find the lightest acceptable W section for $F_y = 36$ ksi.

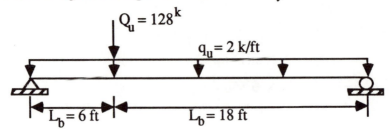

5.10 Find the lightest acceptable W section for $F_y = 36$ ksi. Lateral braces are provided only at the supports and at the cantilever tip.

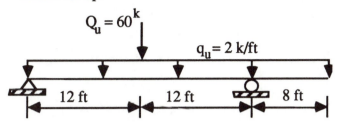

$$Q_u = 60^k$$

$$q_u = 2 \text{ k/ft}$$

12 ft | 12 ft | 8 ft

5.11 $F_y = 50$ ksi
Is a W24X76 acceptable for the loading and L_b values shown in Figure P5.9?

Q_u is due to the interior reaction of a W27X84 sitting on the top flange of the W24X76.

Is a bearing plate required to protect the web of the W27X84? If the answer is yes, find the minimum acceptable value of N.

Is a bearing plate required to protect the web of the W24X76? If the answer is yes, find the minimum acceptable value of N.

Each support for the W24X76 is an 8 inch thick concrete wall of $f'_c = 3$ ksi. For $N = 6$ inches, use the recommended design procedure on LRFD page 3-49 and design the bearing plate for the left reaction.

5.12 C12X20.7 $F_y = 50$ ksi; $L_b = 8$ ft.

In Figure P5.10, Q_u is applied through the shear center. Ignore the beam weight.
Find the maximum acceptable value of Q_u.

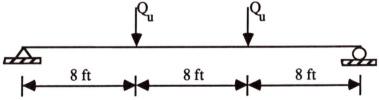

Q_u

Q_u

8 ft | 8 ft | 8 ft

5.13 $F_y = 50$ ksi

Given: The plastic centroid is at 10 in. from the bottom of the section and $\phi_b M_{px} = 470$ ftk.

$J = 6.50$ in^4; $C_w = 1622$ in^4

The shear center is at 2.10 in. from the top of the section.

The elastic centroid is at $\bar{y} = 7.08$ in. from the top of the section.

$r_y = 1.86$ in.

For the compression flange, the elastic section modulus is: $S_{xc} = I_x / \bar{y}$

For the tension flange, the elastic section modulus is:

$$S_{xt} = I_x / (16.25 - \bar{y}).$$

The three plates are continuously and adequately fillet welded along the member length.

Is the design acceptable for the loading shown in Figure P5.13?

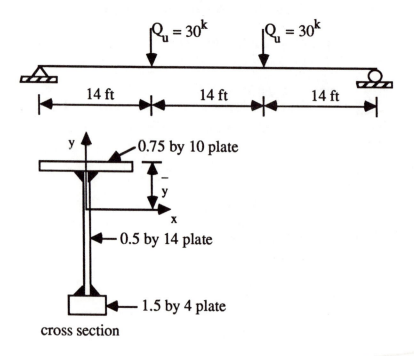

cross section

5.14 $F_y = 36$ ksi

Q_{ux} and Q_{uy} in Figure P5.14 are located at midspan of a simply supported W24X131.

$L_b = L = 30$ ft.

Is the design acceptable?

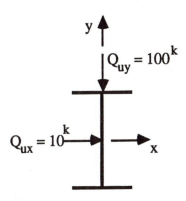

5.15 $F_y = 36$ ksi

Solve problem 5.14 for W30X116.

5.16 $F_y = 50$ ksi

Solve problem 5.14 for W27X94.

5.17 $F_y = 36$ ksi

The Z section loaded as shown in Figure P5.17 is used as a simply supported beam; $L_b = L = 12$ ft. Does the Z section satisfy the LRFD Specs. for bending strength requirements?

Given information:

The shear center and plastic centroid coincide with the elastic centroid.

For the major principal axis(u axis): $\phi_b M_{pu} = 37.1$ ftk

For the minor principal axis(v axis): $\phi M_{pv} = 11.2$ ftk

Verify the following properties:

$A = 6.00$ in^2 ; $I_X = 31.75$ in^4; $I_Y = 11.5$ in^4; $I_{XY} = -14.44$ in^4

$I_u = 39.3$ in^4; $I_v = 3.99$ in^4; $r_v = 0.815$ in.

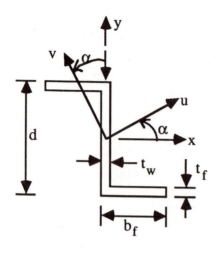

$b_f = 3.50$ in.; $t_f = 0.500$ in.
$d = 6.00$ in.; $t_w = 0.500$ in.

$h = d - t_f = 5.50$ in.

$b = b_f - \dfrac{t_w}{2} = 3.25$ in.

$t = t_f = t_w = 0.500$ in.

$J = \dfrac{t^3(2b+h)}{3}$

$C_w = \dfrac{tb^3h^2}{12}\dfrac{(b+2h)}{(2b+h)}$

5.18 $F_y = 36$ ksi

The Z section in problem 5.17 is a simply supported on rafters and loaded as shown in Figure 5.18. The rafters are 18 ft. on center; that is, $L_b = L = 18$ ft. for the Z section.
Does the Z section satisfy the LRFD Specs. bending strength requirements?

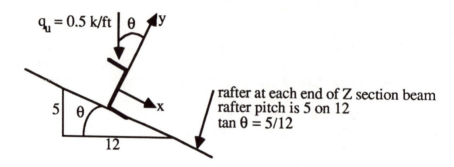

rafter at each end of Z section beam
rafter pitch is 5 on 12
$\tan \theta = 5/12$

5.19 Solve problem 5.18 for 1 on 2 pitch and $L_b = L = 20$ ft.

5.20 Solve problem 5.19 for $F_y = 50$ ksi and $q_u = 0.7$ k/ft.

Chapter 6

MEMBERS SUBJECTED TO AN AXIAL FORCE AND BENDING

6.1 INTRODUCTION

The purpose of this chapter is to discuss the behavior and design of members in frames for which LRFD C1, C2(page 6-35) and LRFD H1, H2(pages 6-47 to 6-49) are applicable.

Figure 6.1 is an example of an industrial building during the early stages of construction. After the steel framework is erected, a flat roof, sidewalls, and roll up doors on the other two sides of the building are to be installed. The roll up doors serve as the endwalls and permit vehicles to enter the building. For discussion purposes, assume that the diagonal members are either single angles or threaded rods and all other members are W sections. The W section members have a strong bending strength axis and a weak bending strength axis. In order to span across the roll up door openings and to resist lateral forces due to wind on the sidewalls, the strong bending strength axis of the vertical and horizontal members in Figure 6.1b is chosen as the axis of bending in the unbraced frame. Consequently, a fully moment resistant connection of the vertical and horizontal members in Figure 6.1b is provided. In the braced frame shown in Figure 6.1a, the weak bending strength axis of the vertical members is the axis of bending. Since the weak axis bending strength of a W section is small, all members in Figure 6.1a are connected such that a negligible moment is transferred between the members at a joint in a braced frame. Therefore, in the frame whose members are pin connected, diagonal members are provided to brace(stabilize) the frame and to resist the wind force on the endwalls(roll up doors). The diagonal members are either single angles or threaded rods which are strong in tension and weak in compression. When the wind force direction is as shown in Figure 6.1a, member 1 is in compression and its buckling strength is small; therefore, member 1 is assumed to be inactive when the structural analysis due to wind is performed. When the wind force direction in Figure 6.1a is reversed, member 1 is in tension and member 2 is assumed to be inactive in the structural analysis due to wind.

Consider the structure represented in Figure 6.1 and the load combinations shown on LRFD page 6-25. Due to wind there is a pressure on the windward side of the building and a suction on the leeward wall and on the flat roof. Suction on the roof is an upward load whereas gravity loads(dead and live) on the roof are downward. All of the members in the unbraced frame and the roof member in the braced frame are subjected to an axial force and bending due to the load combinations shown on LRFD page 6-25. The load combinations of most concern to the structural engineer almost always cause bending and an axial compressive force to occur in these members. Consequently, in the interest of brevity, <u>a member subjected to *bending and an axial compressive force*</u> is referred to as a *beam-column*.

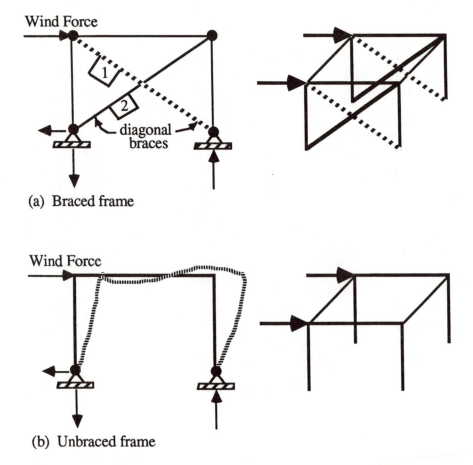

Wind Force

diagonal braces

(a) Braced frame

Wind Force

(b) Unbraced frame

Figure 6.1 Braced and unbraced frames

6.2 SECOND ORDER EFFECTS

Figure 6.2a shows a braced frame for which the exterior walls are attached to girts and the exterior walls are subjected to factored wind loads   and . The length direction of the girts is perpendicular to the plane of Figure 6.1a and the girts are attached to the vertical members at intervals of 4 ft., for example. Consider the load combination defined in LRFD Eqn(A4-6) on LRFD page 6-25 assuming that 1.3W is greater than 0.9D. On the roof, the factored wind load is upward and the factored dead load is downward. For the wind direction shown, member 1 would be inactive(would not resist any load). Member 3 is subjected to an axial tensile force and bending. Members 4 and 5 are beam- columns(members subjected to bending and an axial compression force).

Figure 6.2b shows a braced frame for which the exterior wall panels are attached to beams. The beams span perpendicularly to the plane of Figure 6.2b and are located only at the pinned joint locations. The load combination in Figure 6.2b is assumed to be as defined either in LRFD Eqn(A4-3) or LRFD Eqn(A4-4). In each of these cases, the factored wind load on the roof is upward (due to suction) and the other factored loads are downward (due to gravity). Therefore, for the load direction shown on member 4 in Figure 6.2b, the assumption is that the sum of the factored gravity direction loads exceeds the factored wind load. Member 1 is inactive for the wind direction shown in Figure 6.2b and Member 4 is a beam-column.

In Figure 6.2a, member 3 is subjected to an axial tension force and bending. At each point along the member length in Figure 6.3, the deflection and bending moment are decreased due to the effects of the axial tension force. Therefore, it is conservative to ignore the secondary effects on deflection and bending moment when the design requirements for bending and deflection are checked for such a member.

In Figure 6.2b, member 4 is a beam-column. At each point along the member length in Figure 6.4, the deflection and bending moment are increased due to the effects of the axial compression force. Usually, the secondary effects on deflection and bending moment cannot be ignored when the design requirements for bending and deflection are checked for such a member.

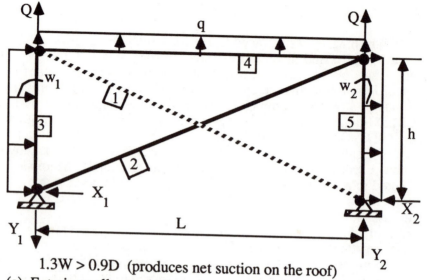

1.3W > 0.9D (produces net suction on the roof)
(a) Exterior walls attached to girts on members 3 and 5

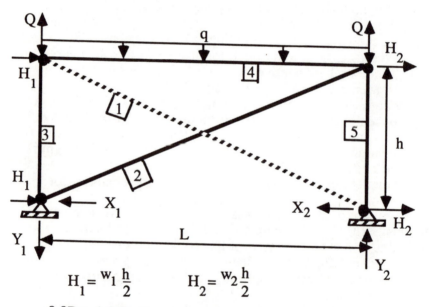

$$H_1 = w_1 \frac{h}{2} \qquad H_2 = w_2 \frac{h}{2}$$

0.9D > 1.3W (does not produce net suction on roof)
(b) Exterior walls attached to beams at hinge locations

Figure 6.2 Braced frame for LRFD Eqn(A4-6) page 6-25

On pages 15 and 29 of reference 6, it is shown for elastic behavior that:

$$v_c \approx \frac{v_{co}}{1 - \rho}$$

where: $\rho = \dfrac{P}{P_e}$

$$P_e = \frac{\pi^2 EI}{L^2}$$

I and L are for the axis of bending
v_{co} is the maximum deflection when P = 0
v_c is the maximum deflection when P > 0

The maximum secondary bending moment(see Figure 6.4c) is:

$$M_s = v_c*P$$

and the maximum total bending moment(seeFigure 6.4d) is:

$$\text{max. } M = \left(\frac{qL^2}{8} + v_c*P\right) \approx \frac{qL^2/8}{1 - \rho}$$

Due to other boundary conditions or/and other types of loads in Figure 6.4a, the maximum total bending moment for factored loads on the member can be obtained by using the information on LRFD page 6-163 in LRFD Eqn(H1-3)(page 6-48) and $M_{LT} = 0$ in LRFD Eqn (H1-2). That is, the maximum total bending moment due to factored loads when the member is not pinned ended is:

$$M_u = \frac{C_m}{1 - \dfrac{P_u}{P_e}} M_{NT}$$

where: $C_m = 1 + \psi\, P_u/P_e$ (see LRFD page 6-162)

ψ is given on LRFD page 6-163
P_u is the required axial compressive strength

$P_e = \pi^2 EI/(KL)^2$

I and KL are for the axis of bending
K is determined from LRFD page 6-151
M_{NT} is the maximum primary bending moment for the case

on LRFD page 6-163 for which ψ was obtained; for the fifth case on LRFD page 6-163, $M_{NT} = 3PL/16$ is obtained from Case 13 on LRFD page 3-134.

NOTE: In lieu of the above definition of C_m, the C_m values given in ii on LRFD page 6-49 may be used.

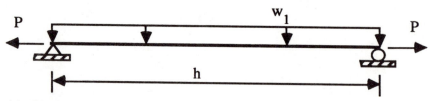

(a) Member 3 of Figure 6.2a

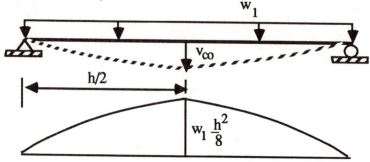

(b) Primary moment diagram is for $P = 0$

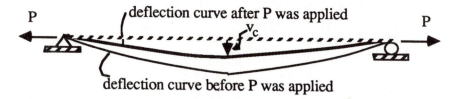

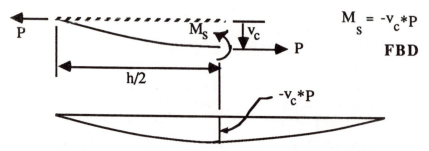

$$M_s = -v_c * P$$

FBD

(c) Secondary moment diagram is P*(final deflected shape)

(d) Combined moment diagram = (b) + (c); max. $M = w_1 \dfrac{h^2}{8} - v_c * P$

Figure 6.3 Secondary effects due to an axial tension force

208

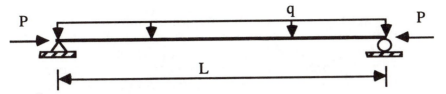

(a) Member 4 of Figure 6.2b

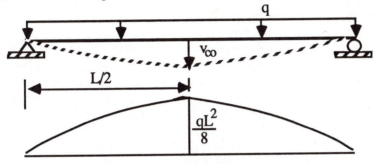

(b) Primary moment diagram is for $P = 0$

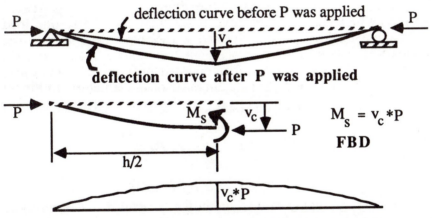

(c) Secondary moment diagram is P*(final deflected shape)

(d) Combined moment diagram = (b) + (c); max. $M = \dfrac{qL^2}{8} + v_c{*}P$

max. $M \approx \dfrac{qL^2/8}{1-\rho}$ where $\rho = \dfrac{P}{P_e}$ and $P_e = \dfrac{\pi^2 EI}{(KL)^2}$

Figure 6.4 Secondary effects due to an axial compression force

Figure 6.5a shows an unbraced frame subjected to a load combination. The reaction directions assumed in Figure 6.5a are typical for the structure and load combination shown. Note that all of the members are beam-columns. In this text, the design basis(see LRFD A5.1 page 6-26) is for the required strength due to factored loads obtained from an elastic structural analysis. Consequently, superposition is valid.

Figure 6.5b shows the primary moment diagram due only to the loads which do not cause any joint translations to occur. Figure 6.5c shows the primary moment diagram due to the loads which cause joint translations to occur. Figure 6.5d shows the secondary effects on deflection and bending moment. At any point on the structure in Figure 6.5a, the combined bending moment is the algebraic sum of the moments at that point in the Figure 6.5 moment diagrams.

It must be noted that most of the available computer analysis software packages perform only first order analyses. That is, they do not compute the secondary effects on deflection and bending moment. In a first order analysis, the solution for all of the needed load combinations can be obtained in one computer run.

There are only a few available computer analysis software packages which perform a second order analysis. Only one load combination at a time can be performed in a second order analysis. The secondary effects on joint displacements(rotations and translations), member end forces, and member end stiffness coefficients are a function of the axial force in each member of the structure. In Figure 6.5d, $\Sigma M_1 = 0$ with the applied joint loads located at the deflected joint positions gives:

$$Y_2 = u*2*(Q + qL/2)/L$$

and $\Sigma F_Y = 0$ gives:

$$Y_1 = 2*(Q + qL/2) - Y_2 = 2*(Q + qL/2)*(1 - u/L).$$

Since the <u>correct value</u> of **u** <u>is unknown</u>, a ***second order analysis is an iterative process***. In the first iteration cycle, $u = u_0$ is assumed and u_1 is obtained from a structural analysis in which the member end stiffness coefficients are a function of the axial force in the member and the system equilibrium equations are for the deflected structure as shown above in the computation of Y_2. If $u_1/u_0 > 1.03$, for example, another iteration cycle is performed assuming that $u = u_1$ and u_2 is obtained from a structural analysis. If $u_2/u_1 \leq 1.03$, convergence has occurred and $u = u_2$; otherwise, the iterative process is

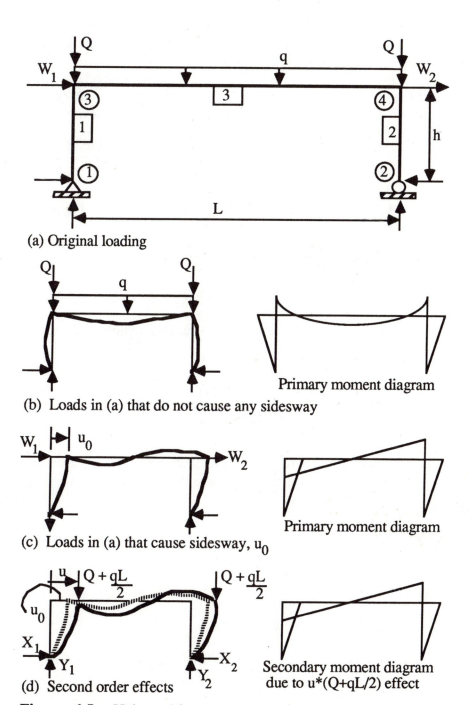

(a) Original loading

Primary moment diagram

(b) Loads in (a) that do not cause any sidesway

Primary moment diagram

(c) Loads in (a) that cause sidesway, u_0

Secondary moment diagram due to $u*(Q+qL/2)$ effect

(d) Second order effects

Figure 6.5 Unbraced frame

211

continued until convergence has occurred and **u** is the computed value obtained in the converged solution.

Since only a few second order structural analysis software packages are available, LRFD H1.2a(LRFD page 6-48) gives a procedure which can be used in lieu of a second order analysis. This procedure applied to Figure 6.5 is:

for each member, the *maximum total moment* due to factored loads is:

$$M_u = B_1 M_{NT} + B_2 M_{LT}$$

where: M_{NT} is the maximum moment for each member in Figure 6.5b

$$\left[B_1 = \frac{C_m}{1 - \dfrac{P_u}{P_e}} \right] \geq 1$$

C_m, P_u, P_e are as defined on page **6-3** in the C_m definition

M_{LT} is the maximum moment for each member in Fig. 6.5c

$$B_2 = \frac{1}{1 - \dfrac{\Sigma P_u}{\Sigma P_e}}$$

$\Sigma P_u = Y_1 + Y_2$ in Figure 6.5d

$\Sigma P_e = \Sigma[\pi^2 EI/(KL)^2]$ for members 1 and 2 in Figure 6.5d

K is obtained from the nomograph on LRFD page 6-153; also see text page 4-7.

Alternatively, $B_2 = \dfrac{1}{1 - \dfrac{\Sigma P_u}{\Sigma H}\left(\dfrac{\Delta_{oh}}{h}\right)}$

where: $\Sigma H = W_1 + W_2$ in Figure 6.5c

Δ_{oh} is u_0 in Figure 6.5c

Alternatively, $B_2 M_{LT}$ may be determined by using the procedure given in the P-Delta Method[9].

6.3 LRFD INTERACTION EQUATIONS

The design requirement for a W section subjected to an axial force and bending is given in LRFD H1(LRFD page 6-47):

$$\text{for } \frac{P_u}{\phi P_n} < 0.2: \quad \frac{P_u}{2\phi P_n} + \left(\frac{M_{ux}}{\phi_b M_{nx}} + \frac{M_{ux}}{\phi_b M_{nx}}\right) \leq 1.00$$

$$\text{for } \frac{P_u}{\phi P_n} \geq 0.2: \quad \frac{P_u}{\phi P_n} + \frac{8}{9}\left(\frac{M_{ux}}{\phi_b M_{nx}} + \frac{M_{ux}}{\phi_b M_{nx}}\right) \leq 1.00$$

where: P_u is the required axial strength

If P_u is a tension force,

ϕP_n is the design tensile strength, $\phi P_n = [\phi_t P_n = 0.9 P_n]$, and P_n is the nominal tensile strength defined in LRFD Chapter D(page 6-36).

If P_u is a compression force,

ϕP_n is the design compression strength,

$\phi P_n = [\phi_c P_n = 0.85 P_n]$,
and P_n is the nominal compressive strength defined in LRFD Chapter E(page 6-39).

$\phi_b M_{nx} = 0.9 M_{nx}$ is the x-axis design bending strength defined in LRFD Chapter F(page 6-42).

$\phi_b M_{ny} = 0.9 M_{nx}$ is the y-axis design bending strength defined in LRFD F1.7(pages 6-45 and 6-92).

M_{ux} is the required x-axis bending strength
M_{uy} is the required y-axis bending strength

If P_u is a tension force, M_{ux} and M_{uy} can be obtained directly from an elastic first order structural analysis.

If P_u is a compression force, M_{ux} and M_{uy} must be obtained from a second order analysis or from the definition of M_u given in LRFD H1.2a(page 6-48) or from the P-Delta Method[9]. NOTE: If the structure is only one member, M_u may be determined as shown on page 207.

The design requirement for shear is given in LRFD F2(page 6-45) and, if P_u is a compression force, LRFD K1.7(page 6-79).

213

In the previous chapters of this text, a member was assumed to be subjected either to an axial force only or to bending only. A beam-column is subjected to an axial compression force and bending. A brief summary review of the design compressive strength and the design bending strength for a W section is appropriate before example beam-column problems are presented.

Consider a W section used as a beam-column. See the limiting width-thickness ratios for compression elements on LRFD page 6-32.

Local buckling is prevented in a W section column when:

$$\frac{0.5b_f}{t_f} \leq \left[\lambda_r = \frac{95}{\sqrt{F_y}}\right] \quad \text{and} \quad \frac{h_c}{t_w} \leq \left[\lambda_r = \frac{253}{\sqrt{F_y}}\right].$$

When local buckling of a column is prevented, the design compressive strength is $\phi_c P_n = 0.85 A_g F_{cr}$ and F_{cr} is as defined on LRFD page 6-39. If local buckling of a column is not prevented, F_{cr} is as defined in LRFD Appendix B(page 6-89). In the first steel design course, the author does not discuss LRFD Appendix B nor does he allow his students to use any W section for which the design compressive strength is defined by LRFD Appendix B. Furthermore, if a W section is used as a beam-column in the first steel design course, the author does not allow his students to use any W section whose flange or/and web width-thickness ratio exceed the λ_p parameter cited in the fourth paragraph below.

The effective length, KL, for each principal axis of a column is needed to determine the compressive design strength. KL is the chord distance between two adjacent M = 0 points on the buckled shape. An M = 0 condition occurs at a real hinge and at a point of inflection on the buckled shape. Each principal axis has a slenderness ratio, KL/r, where r is the radius of gyration for the principal axis associated with KL. When column buckling occurs, the member bends about the principal axis having the larger slenderness ratio. If KL and r are different for each principal axis, KL/r must be computed for both principal axes to determine which axis has the larger KL/r value.

Table C-C2.1 on LRFD page 6-151 gives K values for an isolated column having different boundary conditions. In an unbraced frame, lateral stability depends upon the bending stiffness of the connected beams and columns. For design purposes, the nomograph and the definition of G on LRFD page 6-153 are conservative and can be used to determine K for a column in an unbraced frame provided the assumptions stated on LRFD page 6-152 and amended on text page 4-7 are not violated. However, for an actual building failure investigation, the definition of G given in Eqn(4.16) on text page 4-7 should be used.

In a braced frame, lateral stability is provided by diagonal bracing, shear walls or equivalent means. For a column in a braced frame, K=1 can be conservatively chosen or the "sidesway inhibited" nomograph on LRFD page 2-5 can be used to determine K. A very important assumption not stated on LRFD page 6-152 is that in each of these nomographs, the members perpendicular to the columns do not have any appreciable axial compression in them.

Tabulated values of design compressive strength are given in LRFD pages 2-16 to 2-99 for sections frequently used as a column. In these column tables for W sections, the W14X43(F_y = 50 ksi) is the only entry for which local buckling of a column is not prevented.

For a W section used as a beam-column, if the limiting width-thickness ratio parameter, λ_p , on LRFD page 6-32 is not exceeded for the flanges[$\lambda_p = 65/R(F_y)$] and for the web in combined flexural and axial compression[λ_p is a function of P_u], the strong axis design bending strength, $\phi_b M_{nx}$, is defined in LRFD F1.2 to F1.4(LRFD pages 6-42 to 6-44). Provided the flange width-thickness ratio does not exceed $\lambda_p = 65/R(F_y)$, the weak axis design bending strength is $\phi_b M_{ny} = 0.9Z_y F_y$. If the width-thickness ratio of the flanges or of the web exceed λ_p given on LRFD page 6-32, the design bending strength is defined in LRFD Appendix F(page 6-92). However, in the first steel design course, the author does not discuss LRFD Appendix F nor does he allow his students to use sections for which LRFD Appendix F must be used to determine the design bending strength.

Beam charts, x-axis design bending strength vs. unbraced length curves, for $C_b = 1$ are given in LRFD pages 3-57 to 3-102 for W and M sections. For $1 < C_b \le 2.3$, due to the author's restriction cited in the preceding paragraph, the x-axis design bending strength is the smaller of:

$\phi_b M_{px}$ (LRFD page 3-13)

$C_b*[\phi_b M_{nx}$ obtained from the $C_b = 1$ beam charts(LRFD page 3-57]

6.4 EXAMPLE PROBLEMS

In Examples 6.1 to 6.3, an isolated beam-column is chosen for simplicity reasons to illustrate the basic concepts involved in a member whose ends can not translate perpendicularly to the member axis. However, these basic concepts are applicable to a floor member extracted from a multistory braced frame(see Figure 6.6) at floor level i, for example. Assume for discussion purposes that Figure 6.6 is only one bay of a multibay section on the interior of a building and that the lateral loads shown must be resisted by only the members shown. Suppose a concrete block partition wall is situated directly above each floor member. The wall only needs to extend barely above the ceiling level. An air gap exists between the bottom of the floor member and the top of the partition wall beneath the floor member. Each floor member is subjected to the uniformly distributed dead weight of the partition wall situated above the floor member. If the diagonal braces are designed to resist only tension, the axial compression force in the floor member at the i-th floor level due to the factored wind loads shown is:

$$P_{u(i)} = \Sigma \, W_j \quad \text{(note: } j = i \text{ to } n)$$

Examples 6.4 and 6.5 deal with the basic concepts involved in performing the LRFD interaction equation design check for a beam-column in an unbraced frame using the results from a first order structural analysis.

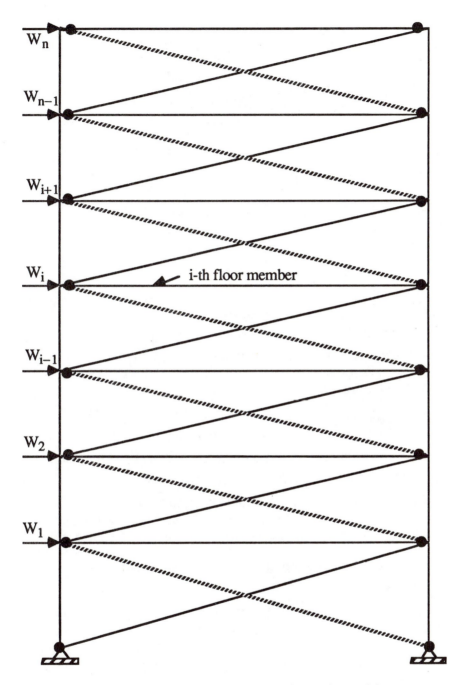

Figure 6.6   One bay of a multibay, multistory braced frame

217

EXAMPLE 6.1_____

The W12X65(F_y = 36 ksi) beam-column in Figure EX6.1 is subjected
to bending about the x-axis only due to the factored distributed load.
Lateral braces are provided only at the member ends. Does the
W12X65 satisfy the requirements of LRFD H1.2(page 6-48)?

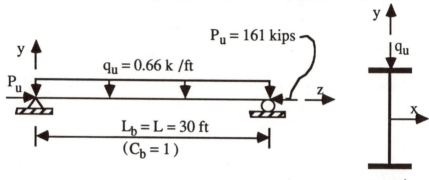

cross section

SOLUTION--See LRFD page 2-24:
Note that a W12X65 F_y = 50 ksi has a flag on it to tell the reader that if

this section is used as a beam-column, $[0.5b_f/t_f] > [\lambda_p= 65/R(50)]$. In
the first steel design course, the author does not allow his students are
to use any W section which has this flag on it.

See LRFD page 2-25:
Note that a W12X40 F_y = 50 ksi has a flag on it to tell the reader that if

this section is used as a beam-column, h_c/t_w must be compared to λ_p
(see the λ_p parameter which is a function of P_u on LRFD page 6-32).

If $h_c/t_w > \lambda_p$, LRFD Appendix F must be used to determine the design
bending strength as a beam-column and the author does not allow his
students in the first steel design course to use LRFD Appendix F.

(EXAMPLE 6.1 solution actually begins here)
As a column:
LRFD page 2-24:
For later purposes, note that a W12X65 F_y = 36 ksi does not have a
flag on it; therefore, the design bending strength of this section as a
beam-column is the same as the design bending strength for a beam.

$$\left[(KL)_Y = 30 \text{ ft}\right] > \left[(KL)_X/(r_x/r_y) = 30/1.75 = 17.14 \text{ ft}\right]$$

$$\phi P_n = [\phi_c P_{ny} = 277 \text{ kips}]$$

As a beam:

The maximum primary moment = $q_u L^2/8 = 0.66*(30)^2/8 = 74.3$ ftk
does not occur at a lateral brace point; therefore, $C_b = 1$.

LRFD page 3-15: $\phi_b M_{px} = 261$ ftk (Note: this was done to get a
rough idea of where to look in the beam-charts for
$\phi_b M_{nx}$.)

LRFD page 3-71: $C_b = 1$; $L_b = 30$ ft; $\phi_b M_{nx} = 212.5$ ftk

As a beam-column:

(see LRFD page 6-48--Eqn(H1-2))
$M_{ux} = B_1*(74.3 \text{ ftk}) + B_2*(0)$

$$\left[B_1 = \frac{C_{mx}}{1 - \dfrac{P_u}{P_{ex}}} \right] \geq 1$$

$$P_{ex} = \pi^2 EI_x /(KL)_x^2 = \pi^2*29000*533/(30*12)^2 = 1177 \text{ kips}$$

$$P_{ex} = \pi^2*29000*533/(30*12)^2 = 1177 \text{ kips}$$

$$C_{mx} = 1 \quad (\text{see LRFD page 6-163})$$

$$\left[B_1 = 1.0/(1 - 161/1177) = 1.1585 \right] \geq 1$$

$M_{ux} = 1.1585*74.3 = 86.1$ ftk

LRFD page 6-47:
$[P_u/(\phi P_n) = 161/277 = 0.581] \geq 0.2$
The design must satisfy LRFD Eqn(H1-1a):

$$\left[0.581 + \frac{8}{9}*\frac{86.1}{212.5} = 0.941 \right] \leq 1.00 \qquad \text{OK}$$

Yes, the W12X65 satisfies the requirements of LRFD H1.2 for the
loading and other conditions given in the problem statement.

EXAMPLE 6.2 _____

The W12X65(F_y = 36 ksi) beam-column in Figure EX6.2 is subjected to bending about the x-axis only due to the factored distributed load. Lateral braces are provided such that: L_b = 15 ft; $(KL)_Y$ = 15 ft. Does the W12X65 satisfy the requirements of LRFD H1.2(page 6-48)?

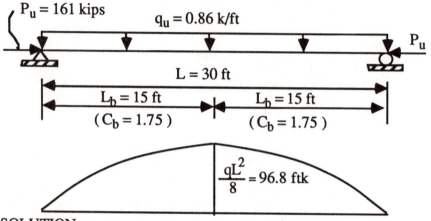

SOLUTION
As a column:
LRFD page 2-24:

$$[(KL)_X/(r_x/r_y) = 30/1.75 = 17.14 \text{ ft}] > [(KL)_Y = 15 \text{ ft}]$$

$\phi P_n = (\phi_c P_{nx} = 458 \text{ kips})$; $[P_u/(\phi P_n) = 261/458 = 0.570] \geq 0.2$

As a beam:
Maximum primary bending moment = $0.86*(30)^2/8 = 96.8$ ftk
LRFD page 3-15: $\phi_b M_{px} = 261$ ftk
LRFD page 3-70, for: $C_b = 1$; $L_b = 15$ ft; $\phi_b M_{nx} = 254.5$
but, $C_b = 1.75$; $(1.75*254.5 = 445) > 261$
     $\phi_b M_{nx} = 261$ ftk

As a beam-column:
$M_{ux} = B_1*(96.8 \text{ ftk}) + B_2*(0)$

$$\left[B_1 = \frac{C_{mx}}{1 - \dfrac{P_u}{P_{ex}}} \right] \geq 1$$

$$P_{ex} = \pi^2 EI_X /(KL)_X^2 = \pi^2 * 29000 * 533/(30*12)^2 = 1177$$

$$\left[B_1 = 1.0/(1 - 261/1177) = 1.285\right] \geq 1$$

$M_{ux} = 1.285*96.8 = 124.4 \text{ ftk}$

$$\left[0.571 + \frac{8}{9} * \frac{124}{261} = 0.994\right] \leq 1.00 \qquad \text{OK (LRFD H1.2 is satisfied)}$$

EXAMPLE 6.3_____

The W12X72(F_y = 50 ksi) beam-column in Figure EX6.3.1 is subjected to bending about the x-axis only due to the factored concentrated load. Lateral braces for both flanges are provided only at the supports and at midspan. Does the W12X72 satisfy the LRFD H1.2(page 6-48) requirements?

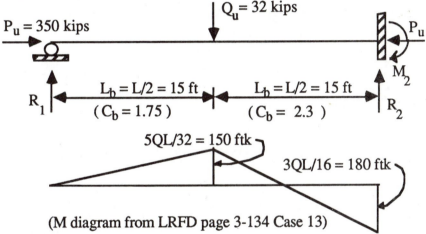

(M diagram from LRFD page 3-134 Case 13)

SOLUTION

<u>As a column:</u>

For y-axis buckling, it is conservative to use $(KL)_Y$ = 15 ft.

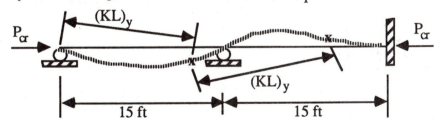

221

For x-axis buckling, see LRFD page 6-151 Case (e): $K_X = 0.8$
$(KL)_X = 0.8*30 = 24$ ft.

LRFD page 2-24:
$[(KL)_Y = 15 \text{ ft}] > [(KL)_X/(r_x/r_y) = 24/1.75 = 13.7 \text{ ft}]$
$\phi P_n = [\phi P_{ny} = 694 \text{ kips}]; \ [P_u/(\phi P_n) = 350/694 = 0.504] \geq 0.2$

<u>As a beam:</u>
The primary moment diagram from LRFD page 3-134 Case 13 is shown above.
For the first $L_b = 15$ ft. region, $C_b = 1.75$.
For the second $L_b = 15$ ft. region:
$[1.75 + 1.05*(150/180) + 0.3*(150/180)^2 = 2.83] > 2.3; \ C_b = 2.3$

LRFD page 3-15: $\phi_b M_{px} = 405$ ftk
LRFD page 3-92:
For $C_b = 1; \ L_b = 15 \text{ft}; \ \phi_b M_{nx} = 393$ ftk
For $C_b = 1.75; L_b = 15 \text{ ft}; (1.75*393 = 688) > 405; \ \phi_b M_{nx} = 405$ ftk
For $C_b = 2.3; \ L_b = 15 \text{ ft}; (2.3*393 = 904) > 405; \ \phi_b M_{nx} = 405$ ftk

<u>As a beam-column:</u>
$P_{ex} = \pi^2*29000*597/(24*12)^2 = 2060$ kips

LRFD page 6-162,163:
$C_m = 1 - 0.3*(P_u/P_{ex}) = 1 - 0.3*350/2060 = 0.949$
$[B_1 = C_{mx}/(1 - P_u/P_{ex}) = 0.949/(1 - 350/2060) = 1.143] \geq 1$
$M_{ux} = 1.143*180 = 206$

$[0.546 + \frac{8}{9}*\frac{206}{405} = 0.998] \leq 1.00$ \hspace{1cm} OK (LRFD H1.2 is satisfied)

EXAMPLE 6.4_____

$F_y = 36$ ksi. for all members(see figure on next page).
Members 1 and 2: W14X48 $L_b = 15$ ft; $(KL)_Y = 15$ ft.
Member 3: W18X40 $L_b = 6$ ft; $(KL)_Y = 6$ ft.

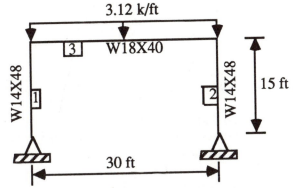

Loading 1: LRFD Eqn(A4-2) with $1.6\,L_r$

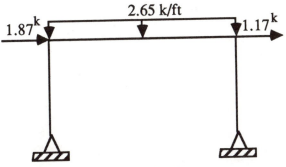

Loading 2: LRFD Eqn(A4-3)

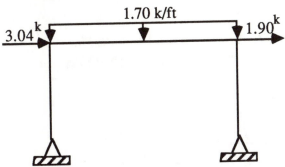

Loading 3: LRFD Eqn(A4-4)

Use the results from an elastic first order analysis for Loadings 1 to 3 in the procedure described in LRFD H1.2(page 6-48). Do the indicated W sections satisfy LRFD H1.2? Author's choice for the maximum drift index is 1/350 for service loads. Check drift control for service loads.

SOLUTION

The FOA(First Order Analysis) results shown below were obtained using the indicated W section properties and load combinations.

LOADING 1: See figure on next page.
No appreciable sidesway deflection occurs due to these loads; $M_{LT} = 0$ in LRFD Eqn(H1-2) page 6-48 and $M_{ux} = B_1 M_{NT}$.
However, sidesway frame buckling is not prevented and ϕP_{crx} due to sidesway frame buckling must be considered.

Check Members 1 and 2: W14X48 $F_y = 36$ ksi

As a column:
For sidesway frame buckling:
LRFD page 6-153:
$G_{bottom} = 10$; $G_{top} = (485/15)/(612/30) = 1.58$; $K_X = 2.01$
$(KL)_X = 2.01 *15 = 30.2$ ft.
LRFD page 2-21:
$$\left[(KL)_Y = 15 \text{ ft}\right] > \left[(KL)_X/(r_x/r_y) = 30.2/3.06 = 9.87 \text{ ft}\right]$$
$\phi P_n = [\phi P_{ny} = 270 \text{ kips}]$

As a beam:
LRFD page 3-16: $\phi_b M_{px} = 212$ ftk
LRFD page 3-72: $C_b = 1$; $L_b = 15$ ft; $\phi_b M_{nx} = 183$ ftk
For $C_b = 1.75$: $(1.75*183 = 320) > 212$; $\phi_b M_{nx} = 212$ ftk

As a beam-column:
$M_{ux} = B_1 M_{NT} = B_1 *(164.6 \text{ ftk})$
$$\left[B_1 = \frac{C_{mx}}{1 - \dfrac{P_u}{P_{ex}}} \right] \geq 1$$
LRFD page 6-48:
The P_{ex} definition for members 1 and 2 is for a braced frame;
use $K_X = 1$; $P_{ex} = \pi^2 EI_X/(KL)_X^2 = \pi^2*29000*485/(180)^2 = 4284$ kips

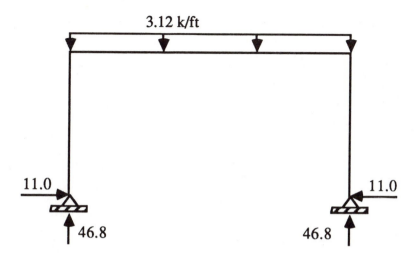

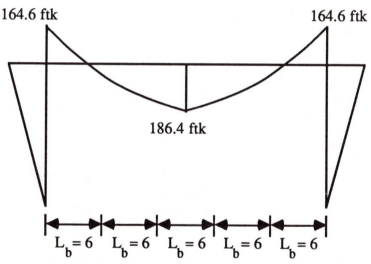

$C_{mx} = 0.6 - 0.4*(0/165) = 0.6$

$[B_1 = 0.6/(1 - 46.8/4284) = 0.607]$; $B_1 \geq 1.00$ is required.

$M_{ux} = = B_1*(164.6 \text{ ftk}) = 1.00*165 = 165 \text{ ftk}$

$[P_u/(\phi P_n) = 46.8/270 = 0.173] < 0.2$

$\left[\dfrac{0.173}{2} + \dfrac{165}{212} = 0.865\right] \leq 1.00$

Members 1 and 2 are OK for Loading 1

Check Member 3: W18X40 $F_y = 36$ ksi
LRFD page 6-32: Check author's restrictions for a beam-column.
$[0.5b_f/t_f = 5.7] \leq [\lambda_r = 65/\sqrt{36} = 10.8]$
        Flange is OK

$[P_u/(\phi P_y) = 11.0/(0.9*A*F_y) = 11.0/(0.9*11.8*36) = 0.0288] \leq 0.125$
$[h_c/t_w = 51.0] \leq [\lambda_r = (1 - 2.75*0.0288)*640/\sqrt{36} = 98.2]$
        Web is OK

As a beam:
$C_b = 1$ (max M = 186 ftk occurs at middle of $L_b = 6$ ft)
LRFD page 3-70: $C_b = 1$; $L_b = 6$ ft; $\phi_b M_{nx} = 206$ ftk

As a beam-column:
LRFD page 6-48; use $K_x = 1$

$P_{ex} = \pi^2*29000*612/(360)^2 = 1352$ kips
Use $C_{mx} = 0.85$ (see item ii on LRFD page 6-49)
$B_1 = 0.85/(1 - 11.0/1352) = 0.86$; $B_1 \geq 1.00$ is required.
$M_{ux} = B_1*(186) = 1.00*186 = 186$ ftk

As a column:

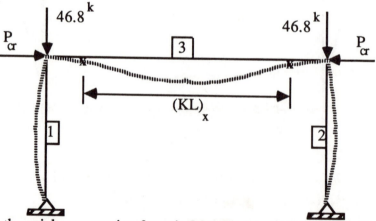

Since the axial compression force in Members 1 and 2 is not negligible, the "sidesway inhibited" nomograph on LRFD page 2-5 cannot be used to obtain K_x for member 3. Be conservative and use $K_x = 1$.
$(KL)_Y = 6$ ft = 72 in. was given in the problem statement.
$(KL/r)_x = 360/7.04 = 51.1$; $(KL/r)_y = 72/1.22 = 59.0$

$$\left[\lambda_{cy} = \frac{(KL)_y}{\pi r_y}\sqrt{\frac{F_y}{E}} = (59.0/\pi)*\sqrt{36/29000} = 0.662\right] \leq 1.5$$

$\phi_c P_{ny} = 0.85*11.8*0.658^{0.662}*36 = 274$ kips
$[P_u/(\phi P_n) = 11.0/274 = 0.0401] < 0.2$

$$\left[\frac{0.0401}{2} + \frac{186}{206} = 0.923\right] \leq 1.00; \quad \text{Member 3 is OK for Loading 1}$$

LOADING 2: (subdivide into no lateral translation solution and lateral translation solution)

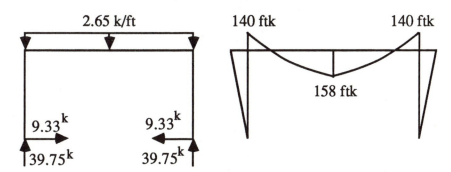

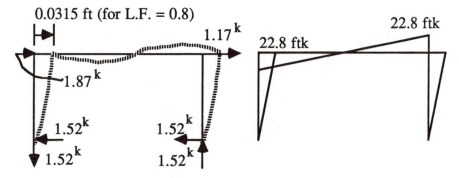

Check drift for service loads:

$$\left[\frac{(0.0315 \text{ ft})/0.8}{15 \text{ ft}} = \frac{1}{381}\right] \leq \frac{1}{350} \quad \text{Drift is OK.}$$

227

Check Members 1 and 2
$\phi P_n = 270$ kips (obtained from Loading 1 Solution)
$\phi_b M_{nx} = 212$ ftk (obtained from Loading 1 Solution)

$$M_{ux} = B_1 M_{NT} + B_2 M_{LT} = B_1*(140 \text{ ftk}) + B_2*(\pm 22.8 \text{ ftk})$$

For the loading component that produces M_{NT}:
$P_{ex} = 4284$ kips (obtained from Loading 1 Solution)
$B_1 = C_{mx}/(1 - P_u/P_{ex}) = 0.6/(1 - 79.6/4284) = 0.611$
$B_1 \geq 1.00$ is required.

For the loading component that produces M_{LT}:
$(KL)_X = 30.2$ ft $= 362$ in. (obtained from Loading 1 Solution)
$P_{ex} = \pi^2*29000*485/(362)^2 = 1059$ kips (for Members 1 and 2)
$$B_2 = \cfrac{1}{1 - \cfrac{\Sigma P_u}{\Sigma P_{ex}}} = \cfrac{1}{1 - \cfrac{79.6}{2118}} = 1.039$$

Alternatively, $$B_2 = \cfrac{1}{1 - \cfrac{\Sigma P_u}{\Sigma H}\left(\cfrac{\Delta_{oh}}{L}\right)} = \cfrac{1}{1 - \cfrac{79.6}{3.04}*\cfrac{0.0315}{15}} = 1.058$$

Member 2 -- Combined results:
$M_{ux} = B_1 M_{NT} + B_2 M_{LT} = B_1*(140 \text{ ftk}) + B_2*(22.8 \text{ ftk})$
$M_{ux} = 1.00*140 + 1.06*22.8 = 164$ ftk
$P_u = B_1*39.8 + B_2*1.52 = 1.00*39.8 + 1.06*1.52 = 41.4$ kips
Member 1 -- Combined results:
$M_{ux} = B_1 M_{NT} + B_2 M_{LT} = B_1*(140 \text{ ftk}) + B_2*(-22.8 \text{ ftk})$
$M_{ux} = 1.00*140 - 1.06*22.8 = 116$ ftk
$P_u = 1.00*39.8 + 1.06*(-1.52) = 38.2$ kips
NOTE: These results for Member 1 were done only for illustration purposes. The author knew by inspection that the combined results for Member 2 were more severe than the combined results for Member 1.

Check Member 2: $[P_u/(\phi P_n) = 41.4/270 = 0.153] \leq 0.2$

$\left[\dfrac{0.153}{2} + \dfrac{164}{212} = 0.850\right] \leq 1.00$; Members 1, 2 are OK for Loading 2.

Member 3 -- Combined results:

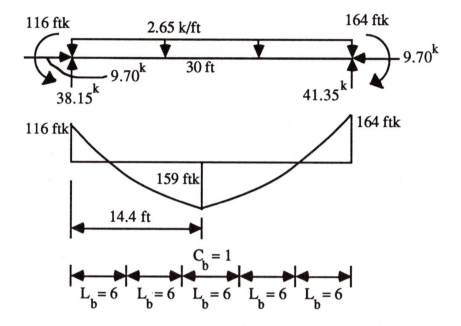

$C_b = 1$; $L_b = 6$ ft; $\phi_b M_{nx} = 206$ ftk (from Loading 1 Solution)

$P_{ex} = 1352$ kips; $\phi P_n = 274$ kips (from Loading 1 Solution)

$[P_u/(\phi P_n) = 9.70/274 = 0.0354] \leq 0.2$

$$\left[\frac{0.0354}{2} + \frac{164}{206} = 0.814 \right] \leq 1.00; \quad \text{Member 3 is OK for Loading 2.}$$

LOADING 3: (subdivide into no lateral translation solution and lateral translation solution)

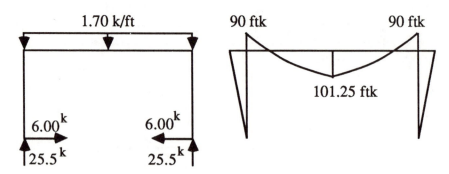

1.70 k/ft

90 ftk   90 ftk

101.25 ftk

6.00^k 6.00^k

25.5^k 25.5^k

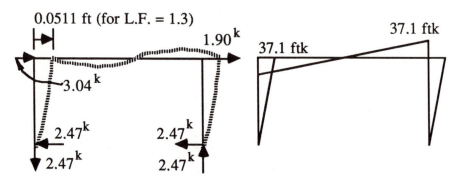

0.0511 ft (for L.F. = 1.3)

1.90^k

37.1 ftk   37.1 ftk

−3.04^k

2.47^k 2.47^k

2.47^k 2.47^k

Check drift for service loads:

$$\left[\frac{(0.0511 \text{ ft})/1.3}{15 \text{ ft}} = \frac{1}{382} \right] \leq \frac{1}{350} \quad \text{Drift is OK.}$$

Check Members 1 and 2
$\phi P_n = 270$ kips (from Loading 1 Solution)
$\phi_b M_{nx} = 212$ ftk (from Loading 1 Solution)

$$M_{ux} = B_1 M_{NT} + B_2 M_{LT} = B_1 *(90 \text{ ftk}) + B_2 *(\pm 37.1 \text{ ftk})$$

For the loading component that produces M_{NT}:
$P_{ex} = 4284$ kips (from Loading 2 Solution)
$B_1 = C_{mx}/(1 - P_u/P_{ex}) = 0.6/(1 - 51.0/4284) = 0.607$
$B_1 \geq 1.00$ is required.

For the loading component that produces M_{LT}:

P_{ex} = 1059 kips (from Loading 2 Solution for Members 1 and 2)

$$B_2 = \cfrac{1}{1 - \cfrac{\Sigma P_u}{\Sigma P_{ex}}} = \cfrac{1}{1 - \cfrac{51.0}{2118}} = 1.025$$

Alternatively, $B_2 = \cfrac{1}{1 - \cfrac{\Sigma P_u}{\Sigma H}\left(\cfrac{\Delta_{oh}}{L}\right)} = \cfrac{1}{1 - \cfrac{51.0}{4.94} * \cfrac{0.0511}{15}} = 1.036$

Member 2 -- Combined results:
M_{ux} = 1.00*90 + 1.04*37.1 = 129 ftk
P_u = 1.00*25.5 + 1.04*2.47 = 28.1 kips
$[P_u/(\phi P_n) = 28.1/270 = 0.104] \leq 0.2$

$$\left[\frac{0.104}{2} + \frac{129}{212} = 0.660\right] \leq 1.00; \quad \text{Members 1, 2 are OK for Loading 3.}$$

Member 3 -- Combined results:

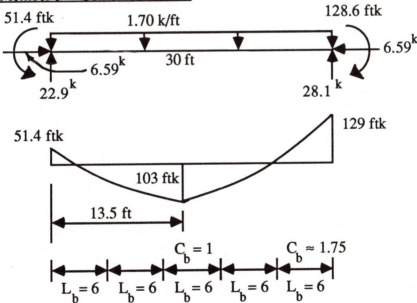

P_{ex} = 1352 kips; ϕP_n = 274 kips (from Loading 1 Solution)
$[P_u/(\phi P_n) = 659/274 = 0.0241] \leq 0.2$

For the $L_b = 6$ ft. region where $C_b = 1$,

$\phi_b M_{nx} = 206$ ftk (from Loading 1 Solution)

$$\left[\frac{0.0241}{2} + \frac{103}{206} = 0.512\right] \leq 1.00 \quad \text{OK}$$

For the $L_b = 6$ ft. region where $C_b \approx 1.75$:

$(1.75*206 = 361) > 212; \quad \phi_b M_{nx} = 212$

$$\left[\frac{0.0241}{2} + \frac{129}{212} = 0.621\right] \leq 1.00 \quad \text{OK}$$

Member 3 is OK for Loading 3.
All members satisfy LRFD H1.2 for all loadings.

EXAMPLE 6.5 _____

Does a W8X28 $F_y = 36$ ksi for Member 1 in Figure 3.2(page **3-5**) satisfy LRFD H1.2? Assume $L_b = 10.5$ ft and $(KL)_Y = 10.5$ ft for Member 1. See text Appendix A for the FOA solutions.

SOLUTION
Lateral stability of the structure in Figure 3.2 is achieved by bending of Members 1 to 4 and their interaction with the truss. To obtain an estimate of $(KL)_X$ for sidesway frame buckling, convert the truss to an equivalent beam(see figure on next page). From text Appendix A:

   Members 5 to 14, A = 5.00 in^2; I = 20.9 in^4
   Members 15 to 24, A = 5.89 in^2; I = 14.4 in^4

For the truss as an equivalent beam, try:

   I = 0.85*[35.3 + (5.00 + 5.89)*(27)] = 6778 in^4

NOTE: The 0.85 factor was chosen by the author based on his experience and accounts for the reduction in stiffness due to shear deformations of a deep beam. The truss is a deep beam with large holes in the web and shear deformations are not negligible. To verify the assumed value of I = 6778, the assumed structure subjected to the loading shown below was analyzed. The FOA lateral deflection results obtained were 2.29 in. which compares very well with the 2.11 in. at joint 3 in text Appendix A for a loading that involves factored lateral wind loads.

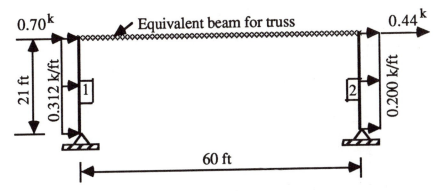

$G_{bottom} = 1$; $G_{top} = (98/279)/(6778/720) = 0.0373$; $K_X = 1.22$

For Member 1: $L = 21$ ft.; $(KL)_X = 1.22*21 = 25.62$ ft. $= 307.4$ in.

See text Appendix A, Member 1:

The maximum axial compression occurs for Loading 7. The maximum moment occurs for Loading 8. Check LRFD H1.2 for these two loadings.

Loading 7: $1.2D + 0.5L + 1.6S$

The lateral deflections due to this loading are negligible.

Member 1: $L = 21$ ft $= 252$ in.

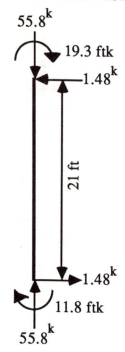

55.8^k

19.3 ftk

-1.48^k

21 ft

1.48^k

11.8 ftk

55.8^k

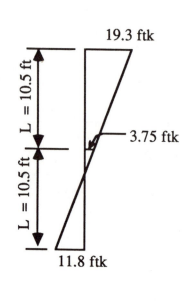

19.3 ftk

3.75 ftk

$L = 10.5$ ft

$L = 10.5$ ft

11.8 ftk

As a braced frame:

$P_{ex} = \pi^2*29000*98/(252)^2 = 442$ kips

$C_m = 0.6 - 0.4*(11.8/19.3) = 0.355$

$B_1 = 0.355/(1 - 55.8/442) = 0.406$

$B_1 \geq 1.00$ is required

LRFD page 2-29:

$\left[(KL)_x/(r_x/r_y) = 25.6/2.13 = 12.0 \text{ ft}\right] > \left[(KL)_x = 10.5 \text{ ft}\right]$

$\phi P_n = [\phi_c P_{nx} = 167 \text{ kips}]$

LRFD page 3-16: $\phi_b M_{px} = 73.4$ ftk; $L_p = 6.8$ ft.

LRFD page 3-75: $C_b = 1$; $L_b = 10.5$ ft; $\phi_b M_{nx} = 68.5$ ftk

See C_b definition on LRFD page 6-43:

For the top $L_b = 10.5$ ft region:

$C_b = 1.75 + 1.05*(-3.75/19.3) + 0.3*(-3.75/19.3)^2 = 1.56$

234

For the bottom $L_b = 10.5$ region:

$C_b = 1.75 + 1.05*(3.75/11.8) + 0.3*(3.75/11.8)^2 = 2.11$
Top $L_b = 10.5$ ft region controls(has larger M_{ux} and smaller C_b)
$(1.56*68.5 = 107) > 73.4$; $\phi_b M_{nx} = 73.4$ ftk

$[P_u/(\phi P_n) = 55.8/167 = 0.334] \geq 0.2$

$\left[0.334 + \dfrac{8}{9} * \dfrac{19.3}{73.4} = 0.568 \right] \leq 1.00$ OK for Loading 7

Loading 8: $1.2D + 0.5L + 0.5S + 1.3W$
Since we are using elastic FOA results modified per the procedure in
LRFD H1.2(page 6-48), we must subdivide Loading 8 using indicated
text Appendix A loadings as follows:
    NLT(No Lateral Translation occurs)
    NLT = 1.2*(Loading 1) + 0.5*(Loading 2) + 0.5*(Loading 3)

    LT(Lateral Translation occurs)
    LT = 1.3*(Loading 4)

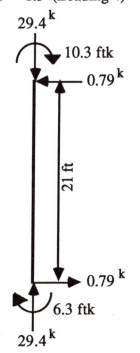

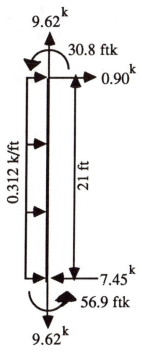

NLT (No Lateral Translation)   **LT** (Lateral Translation)

235

As a braced frame:
$C_m = 0.6 - 0.4*(6.3/10.3) = 0.355$
$P_{ex} = 442$ kips (from Loading 7)
$B_1 = 0.355/(1 - 29.4/442) = 0.380$; $B_1 \geq 1.00$ is required

As an unbraced frame:

$P_{ex} = \pi^2*29000*98/(307)^2 = 298$ kips
$B_2 = 1/(1 - 41.7/596) = 1.075$
Alternatively, $B_2 = 1/[1 - (41.7/13.2)*2.03/252] = 1.026$

$M_{ux} = 1.00*(-6.3) + 1.07*56.9 = 54.6$ ftk
$P_u = 1.00*29.4 + 1.07*(-9.62) = 19.1$ kips

$\phi P_n = 167$ kips (from Loading 7)
$[P_u/(\phi P_n) = 19.1/167 = 0.114] < 0.2$

$$\left[\frac{0.114}{2} + \frac{54.6}{73.4} = 0.801 \right] \leq 1.00 \qquad \text{OK for Loading 8}$$

Check drift deflection due to service load:
From text Appendix Loading 4: $\Delta = 1.64$ in.

$$\text{Drift Index} = \frac{1.64 \text{ in.}}{252 \text{ in.}} = \frac{1}{154}$$

See text page 5-12:
     For this type of building, the author would limit the drift deflection to $h/250 = (252 \text{ in.})/250 = 1.01$ in. Since $[\Delta = 1.64$ in.$] > 1.01$ in., the drift deflection is excessive and a $W8X28(I_x = 98$ in$^4)$ is not adequate for Members 1 to 4 in Figure 3.2 for controlling drift.
     In the next computer analysis, for Members 1 to 4 try a W section for which $I_x \geq \left[(1.64/1.01)*98 = 159 \text{ in}^4 \right]$.
     Try: $W10X33(I_x = 170$ in$^4)$ or W10X39 or W12X40.

EXAMPLE 6.6 _____

The purpose of this example is to show the effect of reducing the drift deflection of the structure shown in Figure EX6.6.1 by:
1. doubling the moment of inertia of the girder only
2. doubling the moment of inertia of the columns only

236

For discussion purposes, the structure shown below is referred to as Structure 1.

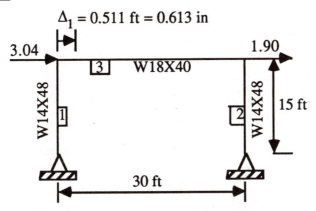

$$\Delta_1 = 0.511 \text{ ft} = 0.613 \text{ in}$$

3.04

1.90

[3] W18X40

W14X48

W14X48

[1]

[2]

15 ft

30 ft

Structure weight, W_1 = 2640 lbs

Structure 1

Structure 2:

Girder: $I_X = 2*612 = 1224 \text{ in}^4$;   Columns: $I_X = 485 \text{ in}^4$

$\Delta_2/\Delta_1 = 0.780$

To obtain $I_X \geq 1224$ for the girder at the least increase in structural steel weight, we would choose:

W24X55 $I_X = 1350$

$W_2/W_1 = (3090 \text{ lbs})/(2640 \text{ lbs}) = 1.17$

If we cannot increase the girder depth, we would choose:

W18X71 $I_X = 1170$

$W_2/W_1 = 3570/2640 = 1.35$

Structure 3:

Columns $I = 2*485 = 970 \text{ in}^4$; Girder $I = 612 \text{ in}^4$

$\Delta_3/\Delta_1 = 0.722$

To obtain $I_X \geq 970$ for the columns, we would choose:

W14X90 $I_X = 999$

$W_3/W_1 = 3900/2640 = 1.48$

If Structure 2 has a W24X55 girder: $W_3/W_2 = 3900/3090 = 1.26$.
If Structure 2 has a W18X71 girder: $W_3/W_2 = 3900/3570 = 1.09$.

237

The author has shown for the chosen example that less structural steel is needed to reduce the drift deflection when the girder moment of inertia is increased. If we added 4 more identical bays to Structure 1, there would be 5 girders and 6 columns for which $W_1 = 10320$ lbs. Changing only the girders from W18X40 to W24X55, $W_2/W_1 = 1.22$, whereas changing from W18X40 to W18X71 girders, $W_2/W_1 = 1.45$. Changing only the columns from W14X84 to W14X90, $W_3/W_1 = 1.37$ (less than $W_2/W_1 = 1.45$ but more than $W_2/W_1 = 1.22$). Thus, the author can not give a general statement that is valid for all structures in regards to whether it is more economical to reduce a drift deflection by increasing the size of the girders or to increase the size of the columns.

6.5 RECOMMENDATIONS FOR FOA TO BE USED IN LRFD EQN(H1-2)

Prior to performing the elastic first order analysis of an unbraced frame, the structural designer performs approximate analyses for gravity loads only and for wind loads only. Approximate solutions for the required LRFD load combinations are obtained by superposing the results found in the approximate gravity only and wind only analyses. The approximate solutions are used in a Preliminary Design procedure to obtain estimated member sizes to be used in the first computer run of the elastic first order analysis. Then, drift deflections are checked; if necessary, some member sizes are increased to control drift and the second computer run is made. When drift deflections are not excessive, the procedure described for LRFD Eqn(H1-2) is used in checking each member in the structure to see if LRFD H1.2 is satisfied. For any member that does not satisfy LRFD H1.2, a new member size is chosen and used in the next computer run. The described design process is continued until LRFD H1.2 is satisfied for all members in the unbraced frame.

In a graduate level steel design course, the author had his students to design an unbraced frame of 18 stories and 3 bays for an office building. The choice of 18 stories was to ensure that excessive story drift deflections due to wind probably would occur unless the students accounted for controlling drift in their initial member sizes based on results from approximate methods of analysis. The FOA results and the procedure described for LRFD Eqn(H1-2) were used in checking LRFD H1.2 for each member. Also, the author's SOA program was used to perform some Second Order Analyses to see how conservative the procedure described for LRFD Eqn(H1-2) was for an 18 story unbraced frame.

Consider Figure 6.7a(page 240) and the application of the procedure described for LRFD Eqn(H1-2) on LRFD page 6-48. A literal interpretation of this procedure for Figure 6.7a requires lateral supports to be provided at each floor level and at the roof level as shown in Figure 6.7b if the gravity loads would cause drift deflections to occur with the lateral supports omitted. Then, in Figure 6.7c the lateral support reactions of Figure 6.7b are reversed and applied along with the lateral loads of Figure 6.7a. For each member in Figure 6.7a, $M_u = B_1*(M_u$ from Figure 6.7b) + $B_2*(M_u$ from Figure 6.7c).

If $B_2 = \dfrac{1}{1 - \dfrac{\Sigma P_u}{\Sigma P_{ex}}}$ is used as the definition of B_2 in the procedure

described for LRFD Eqn(H1-2)[page 6-48], the procedure is the same as the *Moment Magnification method on page 40 of the ACI Code*[10]. Page 56 of the Commentary for the ACI Code states that if the story drift index is not greater than 1/1500 for each story in a structure and loading such as Figure 6.7b with the lateral supports omitted, there is no "appreciable sway" due to the unsymmetric gravity loads and the R forces in Figure 6.7 can be omitted from the analyses. It should be noted that a story drift index of 1/1500 is 1/3 of the maximum out-of-plumb tolerance for steel construction (see LRFD page 6-254) and 2/3 of the maximum out-of-straightness index.

If the R forces in Figure 6.7 can be omitted, an FOA of Figure 6.7 can be made in one computer run. However, if the R forces in Figure 6.7 can not be omitted, two computer runs must be made. In the first computer run, the solution for Figure 6.7b is obtained and the R forces are determined. These R forces are reversed and added onto the lateral loads for the solution of Figure 6.7c in the second computer run. Fortunately, no "appreciable sway" occurs due to the gravity load combinations in many unbraced frames and the solution for all factored loads can be obtained in one computer run. Before the author describes the recommended one-computer-run FOA approach for a multistory unbraced frame, the recommended approach is described for the much simpler case of the one story unbraced frame shown in Figure 6.8.

239

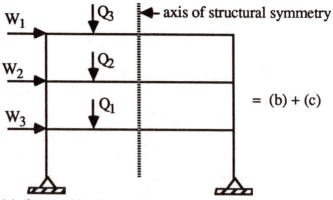

(a) factored load combination

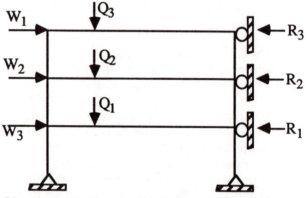

(b) unsymmetric gravity loads produce M_{NT}

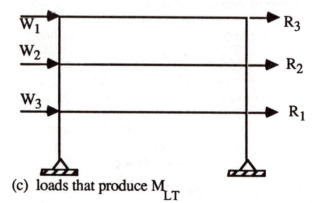

(c) loads that produce M_{LT}

Figure 6.7 FOA for usage in LRFD Eqn(H1-2)

240

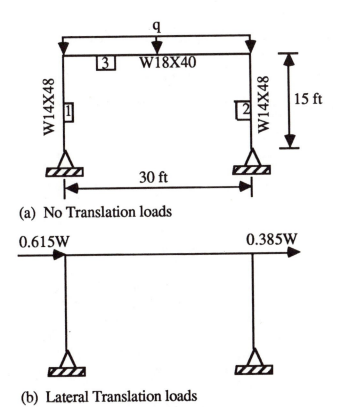

(a) No Translation loads

0.615W →   0.385W →

(b) Lateral Translation loads

Given:
W14X48 columns and W18X40 girder -- A36 steel
Service condition loads:
Gravity -- Dead Load: q = 1.8 k/ft
L_r or S or R: q = 0.6 k/ft
Wind -- q = -0.585 k/ft (minus sign due to suction on roof)
W = 3.80 kips total lateral load at roof level
Maximum drift index = 1/350
Maximum drift deflection = (180 in.)/350 = 0.514 in.

Figure 6.8 One story unbraced frame

The recommended one-computer-run FOA approach for the unbraced frame in Figure 6.8 subjected to the factored load combinations defined by LRFD Eqns(A4-1 to A4-4) is:

241

<u>Step 1</u>
For mathematical convenience, choose q = 1 k/ft and W = 1 kip in the definition of the computer loadings to be superimposed to obtain the factored load combinations required by LRFD Eqns(A4-1 to A4-4) page 6-25.

    Loading 1: q = 1 k/ft in Figure 6.8a
    Loading 2: W = 1 kip in Figure 6.8b

<u>Step 2</u>
Define the one-computer-run factored load combinations in terms of parameters B_1 and B_2 to obtain the results as defined in LRFD Eqn(H1-2).

    LRFD Eqn(A4-1): 1.4**D**
    Loading 3 = B_1*1.4*1.8*(Loading 1)

    LRFD Eqn(A4-2)[author's opinion]: 1.2**D** + 1.6(**L$_r$** or **S** or **R**)
    Loading 4 = B_2*(1.2*1.8 + 1.6*0.6)*(Loading 1)

    LRFD Eqn(A4-3): 1.2**D** + 1.6(**L$_r$** or **S** or **R**) + 0.8**W**
    Loading 5 = (Loading 4) - B_1*(0.8*0.585)*(Loading 1)
              + B_2*(0.8*3.80)*(Loading 2)

    LRFD Eqn(A4-4): 1.2**D** + 0.5(**L$_r$** or **S** or **R**) + 1.3**W**
    Loading 6 = B_1*(1.2*1.8 + 0.5*0.6 - 1.3*0.585)*(Loading 1)
              + B_2*(1.3*3.80)*(Loading 2)

<u>Step 3</u>
Assume values of B_1 and B_2 as described below.

In EXAMPLES 6.4,5, the computed value of B_1 was less than 1. Also, for the 18 story unbraced frame mentioned above, the computed value of B_1 was less than 1 for every member. Consequently, in <u>Loadings 3 to 6 of step 2</u>, assume B_1 = 1.

The maximum acceptable drift index given in Figure 6.8 for service loads is 1/350. Therefore, the maximum acceptable drift deflection <u>for Loading 2 defined in step 2</u> is:
    [(15 ft.)/350]/3.80 = 0.011278 ft. = 0.135 in.

For the loading defined by LRFD Eqn(A4-3), in LRFD Eqn(H1-5)[page 6-49] assume:

$$\frac{\Delta_{oh}}{L} = 0.8 * \frac{1}{350} = \frac{1}{438} \text{ which gives:}$$

$$B_2 = \frac{1}{1 - \frac{\Sigma P_u}{\Sigma H}\left(\frac{\Delta_{oh}}{L}\right)} = \frac{1}{1 - \frac{\Sigma P_u}{\Sigma H}\frac{1}{438}}$$

and since: $\Sigma P_u = (1.2*1.8 + 1.6*0.6 - 0.8*0.585)*30 = 79.6$ kips
$\Sigma H = 0.8*3.80 = 3.04$ kips,

assume: $B_2 = \dfrac{1}{1 - \dfrac{79.6}{3.04}\dfrac{1}{438}} = 1.06$ for Loading 5.

For the loading defined by LRFD Eqn(A4-4), in LRFD Eqn(H1-5)

assume: $\dfrac{\Delta_{oh}}{L} = 1.3 * \dfrac{1}{350} = \dfrac{1}{269}$

and since: $\Sigma P_u = (1.2*1.8 + 0.5*0.6 - 1.3*0.585)*30 = 51.0$ kips
$\Sigma H = 1.3*3.80 = 4.94$ kips,

assume: $B_2 = \dfrac{1}{1 - \dfrac{51.0}{4.94}\dfrac{1}{269}} = 1.04$ for Loading 6.

Step 4

Use assumed values of:

$B_1 = 1$ in Loadings 3 to 6
$B_2 = 1.06$ for Loading 5
$B_2 = 1.04$ for Loading 6

in the computer loadings defined in step 2:

Loading 1: $q = 1$ k/ft in Figure 6.8a
Loading 2: $W = 1$ kip in Figure 6.8b
Loading 3: $2.52*$(Loading 1)
Loading 4: $3.12*$(Loading 1)
Loading 5: $2.65*$(Loading 1) + $3.22*$(Loading 2)
Loading 6: $1.70*$(Loading 1) + $5.14*$(Loading 2)

Results of the one-computer-run FOA for Loadings 5 and 6 are:

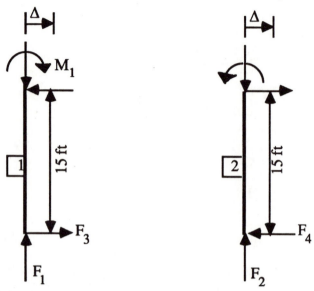

Figure 6.9 Pertinent member end forces and deflection

Loading 5:
$(\Delta = 0.0337$ ft$) \le (1.06*0.8*15/350 = 0.0363$ ft$)$;   drift is OK

$F_1 = 38.1$ kips; $F_2 = 41.4$ kips; $F_3 = 7.71$ kips; $F_4 = 10.9$ kips
$M_1 = 115.6$ ftk; $M_2 = 163.9$ ftk

Check the assumed value of $B_2 = 1.06$
    If the computed value of $B_2 \le 1.06$, the one-computer-run FOA results are usable.

$$\left[B_2 = \frac{1}{1 - \dfrac{\Sigma P_u}{\Sigma H}\left(\dfrac{\Delta_{oh}}{L}\right)} = \frac{1}{1 - \dfrac{79.5}{3.22}\dfrac{0.0337}{15}} = 1.059 \right] \le 1.06 \ \text{OK}$$

    The final design check of LRFD Eqn(H1-2) for Loading 5 is the same as for the combined results of Loading 2 in EXAMPLE 6.4 on page 228 and is not repeated here.

244

<u>Loading 6:</u>
$(\Delta = 0.0534 \text{ ft}) \leq (1.04*1.3*15/350 = 0.0579 \text{ ft});$ drift is OK

$F_1 = 22.9$ kips; $F_2 = 28.1$ kips; $F_3 = 3.41$ kips; $F_4 = 8.55$ kips
$M_1 = 51.1$ ftk; $M_2 = 128.2$ ftk

Check the assumed value of $B_2 = 1.04$
     If the computed value of $B_2 \leq 1.04$, the one-computer-run FOA results are usable.

$$\left[B_2 = \frac{1}{1 - \dfrac{\Sigma P_u}{\Sigma H}\left(\dfrac{\Delta_{oh}}{L}\right)} = \frac{1}{1 - \dfrac{51.0}{5.14}\dfrac{0.0534}{15}} = 1.037 \right] \leq 1.04 \text{ OK}$$

     The final design check of LRFD Eqn(H1-2) for Loading 6 is the same as for the combined results of Loading 3 in EXAMPLE 6.4 on pages 231-232 and is not repeated here.
     It should be noted that if the actual story drift for a loading is less than the maximum acceptable story drift used in computing the assumed B_2 for that loading, the lateral loads are over amplified in the one-computer-run FOA   solution.   If the lateral loads are over amplified, the results from the one-computer-run FOA   are conservative.
     If the R forces in Figure 6.7 can be ignored in a multistory building, the recommended one-computer-run FOA approach described for the load combination defined in LRFD Eqn(A4-4), for example, is:
<u>Step 1</u>.   Choose the maximum acceptable drift index for service loads.
          The author chooses h/400 for illustration purposes where h is the story height.
          Since only the lateral factored wind loads need to be amplifed and L.F. = 1.3 for the chosen load combination, the maximum acceptable drift index for factored loads is 1.3*h/400 = h/308.
<u>Step 2</u>.   In the top story, the middle story, and the bottom story: find an estimated value of

$$B_2 = \frac{1}{1 - \dfrac{\Sigma P_u}{\Sigma H}\left(\dfrac{\Delta_{oh}}{L}\right)} = B_2 = \frac{1}{1 - \dfrac{\Sigma P_u}{\Sigma H}\dfrac{1}{308}}$$

To be conservative, choose the largest estimated B_2 value as the amplification factor for lateral wind loads.   For example, suppose the maximum computed value of $B_2 = 1.12$ and we choose $B_2 = 1.12$ as the amplification factor for lateral wind loads.

Step 3. Assume $B_1 = 1$.

Step 4. LRFD Eqn(A4-4) is:
$1.2D + 0.5L + 0.5(L_r$ or S or $R)] + 1.3W$

For the recommended one-computer-run FOA, the lateral service wind loading is multiplied by $B_2*1.3 = 1.12*1.3 = 1.46$ when the computer loading is assembled. For the loads which were assumed to cause no "appreciable sway" to occur and since we have assumed $B_1 = 1$, the L.F. shown above in LRFD Eqn(A4-4) are the multipliers for these loads when the computer loading is assembled.

Step 5. After the one-computer-run FOA analysis results are available, either use the <u>actual story drift deflection</u> as Δ_{oh} in:

$$B_2 = \cfrac{1}{1 - \cfrac{\Sigma P_u}{\Sigma H}\left(\cfrac{\Delta_{oh}}{L}\right)}$$

or use $B_2 = \cfrac{1}{1 - \cfrac{\Sigma P_u}{\Sigma P_{ex}}}$

and compute B_2 for each story.

If (the computed B_2 for every story) $\leq$ (the assumed $B_2 = 1.12$), the one-computer-run FOA results for the assembled computer loading for LRFD Eqn(A4-4) are usuable provided:

$$\left[B_1 = \frac{C_{mx}}{1 - \frac{P_u}{P_{ex}}} \right] \leq \text{(the assumed } B_1 = 1) \text{ for all members.}$$

Otherwise, another computer run must be made with another set of assumed values for B_1 and B_2. Step 5 is repeated until it is decided that another computer run is not necessary.

6.6 PRELIMINARY DESIGN

Consider the task of sizing the girders and columns in an unbraced multistory plane frame for an office building. See the required LRFD load combinations on LRFD page 6-25. Assume that the building does not have to be designed to resist an earthquake loading.

If there are no more than about 6 stories above ground level in a typical unbraced multistory office building, one of the load combinations which is not a function of wind generally produces the axial force and bending moment values that cause the sum in the applicable interaction equation on LRFD page 6-47 to be a maximum for each member.

The author assumes that each reader has a structural analysis textbook which contains approximate methods of analysis. Therefore, the approximate methods of analysis are not described in this textbook. An approximate method of analysis is performed due to gravity loads only for the factored load combination which is likely to govern most of the member sizes; assume, $B_1 = 1$. Another approximate method of analysis is performed(if there are more than about 6 stories) for the factored load combination with wind which may govern some of the member sizes; assume $B_1 = 1$ for the no "appreciable sway" loads and $B_2 = 1.15$ for the sway loads, for example.

The recommended procedure for selecting a trial W section for a beam-column is:

The author assumes that M_{ux} as defined by LRFD Eqn(H1-2) and P_u have been determined in the approximate analyses; that is, M_{ux} and P_u are known for each approximate analysis for the member being sized.

If it is likely that $\dfrac{P_u}{\phi_c P_n} < 0.2$ when the final design check for strength is made or if the designer wants to use the beam charts to select the trial section, use Recommended Procedure 1 below. Otherwise, use Recommended Procedure 2 below and the column tables to select the trial section.

Recommended Procedure 1 (a beam section is desired)

Convert P_u to an (equivalent M_{ux}) $= P_u * \dfrac{d}{2}$ and assume the desired span/depth ratio is $L/d = 20$ which gives $d = L/20$ and (equivalent M_{ux}) $= \dfrac{P_u L}{40}$.

Use the $C_b = 1$ beam charts(see LRFD page 3-57) and the selection procedure given on page **5-23** to select the lightest W section for which $\phi_b M_{nx} \geq \left[M_{ux} + \dfrac{P_u L}{40} \right]$

where L is the member length.

The selected section is a trial section. Check LRFD Eqn(H1-1b) [or 1a, if applicable]) to see if the trial section is okay.

Recommended Procedure 2 (a column section is desired)

If it is likely that $\dfrac{P_u}{\phi_c P_n} \geq 0.2$ will be applicable when the final design check for strength is made, select the member from the column tables(see LRFD page 2-16). Use the following member selection criterion obtained from LRFD Eqn(H1-1b):

$$\phi_c P_n \geq \left[P_u + \frac{8}{9} * \frac{\phi_c P_n}{\phi_b M_{nx}} * M_{ux} + \frac{8}{9} * \frac{\phi_c P_n}{\phi_b M_{ny}} * M_{uy} \right]$$

$$\phi_c P_n \geq \left[P_u + m_x * M_{ux} + m_y * M_{uy} \right]$$

where: $m_x = \dfrac{8}{9} * \dfrac{\phi_c P_n}{\phi_b M_{nx}}$; $m_y = \dfrac{8}{9} * \dfrac{\phi_c P_n}{\phi_b M_{ny}}$

Convert m_y to an equivalent m_x for later convenience purposes:

$$m_y = \frac{8}{9} * \frac{\phi_c P_n}{\phi_b M_{ny}} * \frac{\phi_b M_{nx}}{\phi_b M_{nx}} = \frac{8}{9} * \frac{\phi_c P_n}{\phi_b M_{nx}} * \frac{\phi_b M_{nx}}{\phi_b M_{ny}}$$

To be conservative, assume: $\phi_b M_{nx} = \phi_b M_{px}$. Therefore, $\dfrac{\phi_b M_{nx}}{\phi_b M_{ny}} = \dfrac{Z_X}{Z_Y}$

where: Z_X and Z_Y are the plastic section moduli given in Part 1 of the LRFD Manual.

248

For the W sections listed in the column tables:

for W14, $\dfrac{Z_X}{Z_Y}$ ranges from 2 to 4;

only the lightest 3 sections exceed $\dfrac{Z_X}{Z_Y} = 3$;

for W12, $\dfrac{Z_X}{Z_Y}$ ranges from 2.20 to 3.42;

only the lightest 3 sections exceed $\dfrac{Z_X}{Z_Y} = 3$;

for W10, $\dfrac{Z_X}{Z_Y}$ ranges from 2.12 to 2.77;

for W8, $\dfrac{Z_X}{Z_Y}$ ranges from 2.15 to 2.16.

Therefore, the author recommends $\dfrac{Z_X}{Z_Y} = 3$ as the assumed value for the first trial section and the selection criterion is:

$$\phi_c P_n \geq \left[P_u + m_x * M_{ux} + 3 * M_{uy} \right]$$

where: $m_x = \dfrac{8}{9} * \dfrac{\phi_c P_n}{\phi_b M_{nx}}$

which can be estimated as shown below.

For $C_b = 1$, $L_b = (KL)_Y$, $\left[(KL)_X / (r_X / r_Y) \right] \leq (KL)_Y$, $F_y = 36$ ksi and $F_y = 50$ ksi:

$\phi_c P_n$ can be obtained from the column tables for a given $(KL)_Y$;

$\phi_b M_{nx}$ can be obtained from the beam charts for $L_b = (KL)_Y$.

Thus, $m_x = \dfrac{8}{9} * \dfrac{\phi_c P_n}{\phi_b M_{nx}}$ was obtained for the W sections listed in the following table.

(KL)$_Y$ (feet)	m$_x$ for F$_y$ = 36 ksi			m$_x$ for F$_y$ = 50 ksi		
	12	16	22	12	16	22
W14X132	1.54	1.46	1.34	1.50	1.41	1.25
W14X48	1.46	1.26	0.90	1.27	1.10	0.80
Recommended	1.5	1.4	1.3	1.4	1.3	1.2
W12X87	1.74	1.61	1.44	1.68	1.55	1.29
W12X40	1.66	1.44	1.05	1.55	1.26	0.92
Recommended	1.7	1.6	1.4	1.6	1.5	1.2
W10X112	1.94	1.76	1.43	1.85	1.62	1.21
W10X33	2.03	1.75	1.27	1.88	1.53	1.12
Recommended	2.0	1.8	1.4	1.9	1.6	1.2
W8X67	2.27	1.93	1.38	2.11	1.70	1.05
W8X31	2.45	2.12	1.54	2.29	1.87	1.21
Recommended	2.3	2.0	1.5	2.2	1.8	1.2

It should be noted that:
1. When $C_b > 1$, m_x is smaller than the tabular values above
2. The recommended $m_x \approx 0.65*m$ for F_y = 36 ksi
$$m_x \approx 0.75*m \text{ for } F_y = 50 \text{ ksi}$$
where **m** is the subsequent approximations value of **m** given on
LRFD page 2-10.

Thus, for the reader's convenience in using LRFD page 2-10, the
recommended selection criterion for the first trial section is:

for F_y = 36 ksi : $\phi_c P_n \geq \left[P_u + 0.65*M_{ux} + 3*M_{uy} \right]$

for F_y = 50 ksi : $\phi_c P_n \geq \left[P_u + 0.75*M_{ux} + 3*M_{uy} \right]$

where: **M$_{ux}$** and **M$_{uy}$** are as defined by LRFD Eqn(H1-2)[page 6-49]
and **m** is the subsequent approximations value of **m** given on
LRFD page 2-10. It should be noted that on LRFD page 2-
10, the $C_m/0.85$ footnote is not applicable since C_m is
correctly used in obtaining M_{ux} and M_{uy} and is not contained
in the definition of **m**.

For a column in a multistory building, it is likely the member is not subjected to any transverse loading and $C_m < 1$. However, the author does not recommend using $C_m < 0.5$ in selecting <u>trial sections based on results obtained from approximate analyses.</u>

EXAMPLE 6.6_____

Given: $P_u = 11.0$ kips; $M_{NT} = 186$ ftk; $M_{LT} = 0$; $C_m = 0.85$
$\quad\quad$ $L = 30$ ft; $C_b = 1$; $L_b = (KL)_Y = 6$ ft; $(KL)_X = 30$ ft.
Select the lightest A36 W section using <u>Recommended Procedure 1</u>.

SOLUTION
Assume: $B_1 = 1$.
$M_{ux} = B_1 * M_{NT} + B_2 * M_{LT} = 1*186 + 0 = 186$ ftk
Convert P_u to an (equivalent M_{ux}) $= P_u L/40$
Try $M_{ux} = \left[M_{ux} + P_u L/40 = 186 + 11.0*30/40 = 194 \text{ ftk} \right]$

LRFD page 3-70: $C_b = 1$; plot point: $[L_b = 6, M_{ux} = 194]$
$\quad\quad\quad\quad\quad\quad$ W16X40 $(\phi_b M_{nx} = 197) \geq 194$
$\quad\quad\quad\quad\quad\quad$ W18X40 $(\phi_b M_{nx} = 206) \geq 194$

This is how the author chose the W18X40 for member 3 in the structure shown on page 223. See page 226 for the design strength check of this trial section where it is shown that $B_1 = 1$ as assumed.

EXAMPLE   6.7_____------

Given: $P_u = 11.0$ kips; $M_{NT} = 186$ ftk; $M_{LT} = 0$; $C_m = 0.85$
$\quad\quad$ $L = 30$ ft; $C_b = 1.75$; $L_b = (KL)_Y = 15$ ft; $(KL)_X = 30$ ft.
Note that only C_b, L_b, and $(KL)_Y$ for this example differ from the given information for EXAMPLE 6.6

Select the lightest A36 W section using recommended Procedure 1.

SOLUTION
Assume: $B_1 = 1$.
$M_{ux} = B_1 * M_{NT} + B_2 * M_{LT} = 1*186 + 0 = 186$ ftk
Convert P_u to an (equivalent M_{ux}) $= P_u L/40$

Try $M_{ux} = [M_{ux} + P_u L/40 = 186 + 11.0*30/40 = 194 \text{ ftk}]$

LRFD page 3-16: W18X40 ($\phi_b M_{px} = 212$) ≥ 194
LRFD page 3-70:
$C_b = 1.75$; plot point [$L_b = 15$, $M_{ux}/C_b = 194/1.75 = 112$]
W18X40 lies to right of and above this point.
Choose W18X40 as the trial section.

Check W18X40:
LRFD page 6-32: Check author's restrictions for a beam-column.
$[0.5b_f/t_f = 5.7] \leq [\lambda_r = \dfrac{65}{\sqrt{F_y}} = 10.8]$ Flange is OK.

$[P_u/(\phi P_y) = 11.0/(0.9*A*F_y) = 11.0/(0.9*11.8*36) = 0.0288] \leq 0.125$
$[h_c/t_w = 51.0] \leq [\lambda_r = (1 - 2.75*0.0288)*640/\sqrt{36} = 98.2]$
Web is OK.

As a beam-column:
$P_{ex} = \pi^2*29000*612/(360)^2 = 1352 \text{ kips}$
$B_1 = 0.85/(1 - 11.0/1352) = 0.86$; $B_1 \geq 1.00$ is required.
$B_1 = 1$ as assumed; therefore, $M_{ux} = 186 \text{ ftk}$ as assumed.

As a beam:
$\phi_b M_{px} = 212 \text{ ftk}$
LRFD page 3-74: $C_b = 1$; $L_b = 15$; $\phi_b M_{nx} = 139 \text{ ftk}$
but $C_b = 1.75$; $(1.75*139 = 243) > 212$; $\phi_b M_{nx} = 212 \text{ ftk}$

As a column:
$(KL/r)_X = 30*12/7.04 = 51.1$; $(KL/r)_Y = 15*12/1.22 = 147.5$

$[\lambda_{cy} = (147.5/\pi)*\sqrt{\dfrac{36}{29000}} = 1.655] > 1.5$

$\lambda_{cy}^2 = 2.738$

$\phi P_n = 0.85*11.8*(0.877/2.738)*36 = 115.7 \text{ kips}$

$[P_u/(\phi P_n) = 11.0/115.7 = 0.0951] < 0.2$

$\left[\dfrac{0.09512}{2} + \dfrac{186}{212} = 0.925\right] \leq 1.00$ OK

EXAMPLE 6.8 _____

Given: $P_u = 261$ kips; $M_{NT} = 96.8$ ftk; $M_{LT} = 0$; $C_m = 1$
$\qquad$ $L = 30$ ft; $C_b = 1.75$; $L_b = (KL)_Y = 15$ ft; $(KL)_X = 30$ ft.
Use <u>Recommended Procedure 2</u> and for A36 steel, select:
$\qquad$ a) lightest W12 listed in the column tables
$\qquad$ b) lightest W14 listed in the column tables.

SOLUTION for lightest W12
Assume: $B_1 = 1$.
$M_{ux} = B_1*M_{NT} + B_2*M_{LT} = 1*96.8 + 0 = 96.8$ ftk

LRFD page 2-10:
Assume: $KL = 15$; $m = 2.4$ for $F_y = 36$ ksi and W12
Try $\phi P_n \geq \left[P_u + 0.65*m*M_{ux} = 261 + 0.65*2.4*96.8 = 412 \text{ kips} \right]$

LRFD page 2-24: $(KL)_Y = 15$; W12X65 $(\phi_c P_{ny} = 485) \geq 412$
$\left[(KL)_X/(r_x/r_y) = 30/1.75 = 17.14 \text{ ft} \right] > \left[(KL)_Y = 15 \right]$
$[\phi P_n = \phi_c P_{nx} = 460 - 0.14*14 = 458 \text{ kips}] \geq 412$

Try W12X65
This is how the author chose the W12X65 for EXAMPLE 6.2
See pages 220-221 for the design check which is:
$\left[0.570 + \frac{8}{9}*\frac{124}{261} = 0.992 \right] \leq 1.00 \qquad$ W12X65 is OK.

SOLUTION for lightest W14
Assume: $B_1 = 1$.
$M_{ux} = B_1*M_{NT} + B_2*M_{LT} = 1*96.8 + 0 = 96.8$ ftk

LRFD page 2-10:
Assume: $KL = 15$; $m = 2.0$ for $F_y = 36$ ksi and W14
Try $\phi P_n \geq [P_u + 0.65*m*M_{ux} = 261 + 0.65*2.0*96.8 = 387 \text{ kips}]$

LRFD page 2-21: $(KL)_Y = 15$; W14X61 $(\phi_c P_{ny} = 412) \geq 387$
$\left[(KL)_X/(r_x/r_y) = 30/2.44 = 12.3 \right] < \left[(KL)_Y = 15 \right]$
$\phi P_n = 412$; $\left[P_u/(\phi P_n) = 261/412 = 0.633 \right] > 0.2$

Try W14X61

$P_{ex} = \pi^2 * 29000 * 640/(360)^2 = 1413.4$ kips
$B_1 = 1/(1 - 261/1413.4) = 1.226$
$M_{ux} = 1.226 * 96.8 + 0 = 118.7$ ftk

LRFD page 3-15: W14X61 $\phi_b M_{px} = 275$ ftk
LRFD page 370: $C_b = 1$; $L_b = 15$; $\phi_b M_{nx} = 257$
but $C_b = 1.75$; $(1.75 * 257 = 450) > 275$; $\phi_b M_{nx} = 275$ ftk

$$\left[0.633 + \frac{8}{9} * \frac{118.7}{275} = 1.017 \right] > 1.00 \quad NG$$

W14X61 is NG, but only violates the interaction equation by 1.7%.
Use W14X68; it obviously is OK.

EXAMPLE 6.9 _____

Given: $P_u = 46.8$ kips; $M_{NT} = 165$ ftk; $M_{LT} = 0$; $C_m = 0.6$
$\quad\quad$ L = 15 ft; $C_b = 1.75$; $L_b = (KL)_Y = 15$ ft; $(KL)_X = 30.2$ ft.

Use <u>Recommended Procedure 2</u> and for A36 steel, select:
$\quad$ a) lightest W14 listed in the column tables
$\quad$ b) lightest W12 listed in the column tables.

SOLUTION for lightest W14
Assume: $B_1 = 1$.
$M_{ux} = B_1 * M_{NT} + B_2 * M_{LT} = 1 * 165 + 0 = 165$ ftk

LRFD page 2-10:
Assume: KL = 15; m = 2.0 for $F_y = 36$ ksi and W14
Try $\phi P_n \geq \left[P_u + 0.65 * m * M_{ux} = 46.8 + 0.65 * 2.0 * 165 = 261 \text{ kips} \right]$

LRFD page 2-21: W14X48 $(\phi_c P_{ny} = 270) \geq 261$
$\left[(KL)_X/(r_x/r_y) = 30.2/3.06 = 9.87 \right] < [(KL)_Y = 15]$
$\phi P_n = 270$; $\left[P_u/(\phi P_n) = 46.8/270 = 0.173 \right] < 0.2$

Try W14X48. This is how the author chose the W14X48 for members
1 and 2 of the structure on page 223 due to Loading 1 which governed
for EXAMPLE 6.4. See text page 225 for the design check which is:

$\left[\dfrac{0.173}{2} + \dfrac{165}{212} = 0.865\right] \leq 1.00$ OK, but W14X43 may work.

Check W14X43:
LRFD page 2-21: W14X43 F_y = 36 ksi has a flag on it.
Must check web requirement on LRFD page 6-32 as a beam-column.
$\left[P_u/(\phi P_n) = 46.8/(0.9*12.6*36) = 0.115\right] \leq \left[\lambda_p = 0.125\right]$
$\left[h_c/t_w = 37.4\right] \leq \left[(640/\sqrt{36})*(1 - 2.75*0.115) = 72.9\right]$
Web is OK.

LRFD page 2-21: $(KL)_Y$ = 15; $\phi_c P_{ny}$ = 239 kips

$\left[(KL)_X/(r_x/r_y) = 30.2/3.08 = 9.81\right] < [(KL)_Y = 15]$
$\phi P_n = 239$; $[P_u/(\phi P_n) = 46.8/239 = 0.196] < 0.2$

$P_{ex} = \pi^2*29000*428/(30.2*12)^2 = 932.7$ kips
$B_1 = 0.6/(1 - 46.8/932.7) = 0.632$; $B_1 \geq 1$ is required.
$M_{ux} = 1.00*165 + 0 = 165$ ftk

LRFD page 3-16: ϕM_{px} = 188 ftk
LRFD page 3-72: C_b = 1; L_b = 15; $\phi_b M_{nx}$ = 160 ftk
but C_b = 1.75; (1.75*160 = 280) > 188; $\phi_b M_{nx}$ = 188 ftk

$\left[\dfrac{0.196}{2} + \dfrac{165}{188} = 0.976\right] \leq 1.00$ OK, use W14X43.

SOLUTION for lightest W12
Assume: B_1 = 1.
$M_{ux} = B_1*M_{NT}+ B_2*M_{LT} = 1*165 + 0 = 165$ ftk

LRFD page 2-10:
Assume: KL = 15; m = 2.4 for F_y = 36 ksi and W12
Try $\phi P_n \geq \left[P_u + 0.65*m*M_{ux} = 46.8 + 0.65*2.4*165 = 304 \text{ kips}\right]$

LRFD page 2-21: W12X53 $(\phi_c P_{ny} = 362) \geq 304$
$\left[(KL)_X/(r_x/r_y) = 30.2/3.07 = 9.84\right] < \left[(KL)_Y = 15\right]$
$\phi P_n = 362$; $[P_u/(\phi P_n) = 46.8/362 = 0.129] < 0.2$

Try W12X53

$P_{ex} = \pi^2*29000*425/(30.2*12)^2 = 1057$ kips

$B_1 = 0.6/(1 - 46.8/1057) = 0.628$; $B_1 = 1.00$ required.

$M_{ux} = 165$ ftk as originally assumed.

LRFD page 3-16: $\phi_b M_{px} = 210$ ftk

LRFD page 3-72: $C_b = 1$; $L_b = 15$; $\phi_b M_{nx} = 197$ ftk

but $C_b = 1.75$; $(1.75*197 = 345) > 210$; $\phi_b M_{nx} = 210$ ftk

$$\left[\frac{0.129}{2} + \frac{165}{210} = 0.851\right] \leq 1.00 \qquad \text{OK, but W12X50 may work.}$$

Check W12X50:

$$\left[\frac{46.8}{2*289} + \frac{165}{195} = 0.927\right] \leq 1.00 \qquad \text{OK, but W12X45 may work.}$$

Check W12X45: $\left[\dfrac{46.8}{2*257} + \dfrac{165}{175} = 1.034\right] > 1.00 \qquad \text{NG}$

Use W12X50.

EXAMPLE 6.10_____

Solve Example 4 on LRFD page 2-11 using author's recommended Procedure 2 and select lightest W14 of A36 steel.

SOLUTION

Assume: $B_1 = 1$ for x-axis bending; $M_{ux} = 1*180 + 0 = 180$ ftk

$\qquad\qquad B_1 = 1$ for y-axis bending; $M_{uy} = 1*60 + 0 = 60$ ftk

LRFD page 2-10: $KL = 14$; $m = 2.0$ for W14 $F_y = 36$ ksi

Try $\phi P_n \geq \left[300 + 0.65*2.0*(180 + 3*60) = 768 \text{ kips} \right]$

LRFD page 2-20: $(KL)_y = 14$; W14X99 $(\phi_c P_{ny} = 799) \geq 768$

Try W14X99: It is OK; the check is shown on LRFD pages 2-11,12.

Try W14X90: $\left[\dfrac{300}{716} + \dfrac{8}{9} * \left(\dfrac{180}{424} + \dfrac{60}{204}\right) = 1.06\right] > 1.00 \quad \text{NG}$

Use W14X99.

Comparison of "first trial selection" procedure results for a beam-column

(Text Example) Lightest W $F_y = 36$ ksi	LRFD Procedure LRFD page 2-10	Author's Recommended Procedures	
		$m_x = 0.65*m$ $Z_X/Z_Y = 3$	m_x recommended on page 250 $Z_X/Z_Y = 3$ initially
(EX 6.8) W12X65	$P_{ueff} = 261+2.4*96.8=493$ try W12X65	$\phi P_n \geq 412$ try W12X65	$\phi P_n \geq [261+1.6*96.8 = 416]$ try W12X65
W14X68	$P_{ueff} = 261+2.0*96.8=455$ try W14X68	$\phi P_n \geq 387$ try W14X61 (interaction eqn. sum = 1.017 for W14X61)	$\phi P_n \geq [261+1.4*96.8 = 397]$ try W14X61
(EX 6.9) W14X43	$P_{ueff} = 46.8+2*165 = 377$ try W14X61	$\phi P_n \geq 261$ try W14X48	$\phi P_n \geq [46.8+1.4*165 = 278]$ try W14X48
W12X50	$P_{ueff} = 46.8+2.4*165=443$ try W12X65(W12X58)	$\phi P_n \geq 304$ try W12X53	$\phi P_n \geq [46.8+1.6*165 = 311]$ try W12X53
(EX 6.10) W14X99	$P_{ueff} = 822$ try W14X99	$\phi P_n \geq 768$ try W14X99	$\phi P_n \geq 804$ try W14X99 $\phi P_n \geq 728$ $(Z_X/Z_Y = 2.1)$ try W14X90 [$\Sigma =1.06$]
W12X106	$P_{ueff} =931[U=1.38]$ try W12X120	$\phi P_n \geq 863$ try W12X120	$\phi P_n \geq 876$ try W12X120 $\phi P_n \geq 798$ $(Z_X/Z_Y = 2.19)$ try W12X106

PROBLEMS

Note to students:

If your professor forgets to specify the grade of steel in problems 6.1 to 6.10, use the information for $F_y = 36$ ksi in checking the design requirement for a beam-column(see LRFD H1.2 page 6-48).

$(KL)_Y = L_b$ is to be used in all problems unless otherwise noted.

6.1 No y-axis bending.

W14X159 $F_y = 36$ ksi   W14X132 $F_y = 50$ ksi

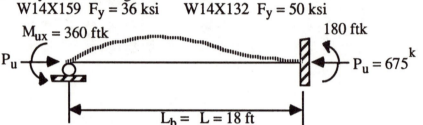

$M_{ux} = 360$ ftk

180 ftk

P_u

$P_u = 675^k$

$L_b = L = 18$ ft

6.2 No y-axis bending.

W14X193 $F_y = 36$ ksi   W14X159 $F_y = 50$ ksi

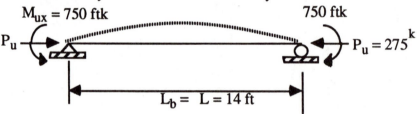

$M_{ux} = 750$ ftk

750 ftk

P_u

$P_u = 275^k$

$L_b = L = 14$ ft

6.3 No y-axis bending.

W14X82 $F_y = 36$ ksi   W14X61 $F_y = 50$ ksi

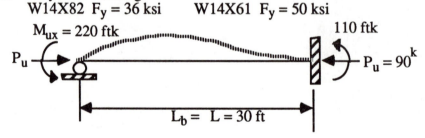

$M_{ux} = 220$ ftk

110 ftk

P_u

$P_u = 90^k$

$L_b = L = 30$ ft

6.4 No y-axis bending.
    W14X90 $F_y = 36$ ksi   W14X74 $F_y = 50$ ksi

$M_{ux} = 130$ ftk

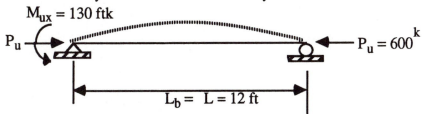

P_u $P_u = 600^k$

$L_b = L = 12$ ft

6.5 No x-axis bending.
    W8X31 $F_y = 36$ ksi   W8X24 $F_y = 50$ ksi

$q_u = 0.8$ k/ft

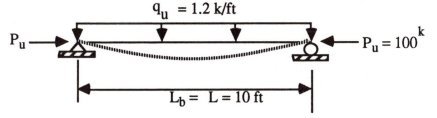

P_u $P_u = 170^k$

$L_b = L = 10$ ft

6.6 No x-axis bending.
    W8X48 $F_y = 36$ ksi   W8X40 $F_y = 50$ ksi

$q_u = 1.2$ k/ft

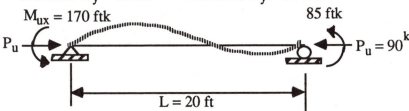

P_u $P_u = 100^k$

$L_b = L = 10$ ft

6.7 No y-axis bending. $(KL)_Y = L_b = 10$ ft.
    W14X61 $F_y = 36$ ksi   W14X48 $F_y = 50$ ksi

$M_{ux} = 170$ ftk   85 ftk

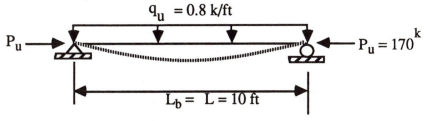

P_u $P_u = 90^k$

$L = 20$ ft

6.8 No y-axis bending. $(KL)_Y = L_b = 8$ ft.
   W12X65 $F_y = 36$ ksi   W12X58 $F_y = 50$ ksi
   $M_{ux} = 140$ ftk

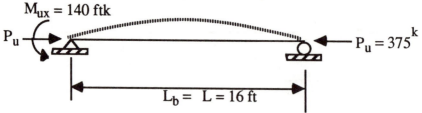

$$L_b = L = 16 \text{ ft}$$

6.9 No y-axis bending.
   W12X136 $F_y = 36$ ksi   W12X120 $F_y = 50$ ksi
   $Q_u = 60^k$

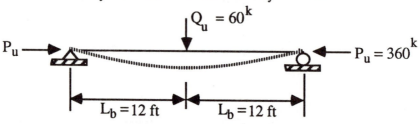

$$L_b = 12 \text{ ft} \qquad L_b = 12 \text{ ft}$$

6.10 No y-axis bending $(KL)_Y = F_y = 15$ ft.
   W12X58 $F_y = 36$ ksi   W12X45 $F_y = 50$ ksi
   $q_u = 1$ k/ft

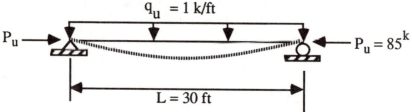

$$L = 30 \text{ ft}$$

In the following problems, use the author's recommended design procedures(see page **6-29**) and find the lightest acceptable W section in the indicated series that satisfies LRFD H1.2(page 6-48).

6.11 See Example 6.1 page **6-9**. W12 $F_y = 50$ ksi

6.12 See Example 6.1 page **6-9**. W14 $F_y = 36$ ksi

6.13 See Example 6.2 page **6-10**. W12 $F_y = 50$ ksi

6.14 See Example 6.2 page **6-10**. W14 $F_y = 36$ ksi

6.15 See Example 6.3 page **6-11**. W12 $F_y = 36$ ksi

6.16 See Example 6.3 page **6-11**. W14 $F_y = 50$ ksi

6.17 See Example 6.4 page **6-13**. $F_y = 50$ ksi
    Try the following:
    Columns: W14X43 and girder: W18X35

6.18 See Example 6.7 page **6-31**. $F_y = 50$ ksi...Any W section.

6.19 See Example 6.10 page **6-35**. W14 $F_y = 50$ ksi

6.20 See Example 6.10 page **6-35**. W12 $F_y = 36$ ksi

Computer Output for an Example of an Elastic, Factored Load Analysis

PROBLEM TITLE : **Structure shown in Figures 3.1 and 3.2**

PLANE FRAME ANALYSIS

Units are: **INCHES, KIPS, RADIANS**

Modulus of Elasticity: **E= 29000**

JOINT NUMBER	JOINT COORDINATES X	Y	Z
1	0.000	0.000	0.000
2	0.000	252.000	0.000
3	0.000	306.000	0.000
4	72.000	252.000	0.000
5	72.000	312.000	0.000
6	144.000	252.000	0.000
7	144.000	318.000	0.000
8	216.000	252.000	0.000
9	216.000	324.000	0.000
10	288.000	252.000	0.000
11	288.000	330.000	0.000
12	360.000	252.000	0.000
13	360.000	336.000	0.000
14	432.000	252.000	0.000
15	432.000	330.000	0.000
16	504.000	252.000	0.000
17	504.000	324.000	0.000
18	576.000	252.000	0.000
19	576.000	318.000	0.000
20	648.000	252.000	0.000
21	648.000	312.000	0.000
22	720.000	252.000	0.000
23	720.000	306.000	0.000
24	720.000	0.000	0.000

MEMBER DATA

Member Number	Incidences Joint 1	Joint 2	AREA	IZ	C1	C2	C3	C4
1	1	2	8.25	98.000	1.0	4.0	2.0	4.0
2	2	3	8.25	98.000	1.0	4.0	2.0	4.0
3	24	22	8.25	98.000	1.0	4.0	2.0	4.0
4	22	23	8.25	98.000	1.0	4.0	2.0	4.0
5	2	4	5.00	20.900	1.0	4.0	2.0	4.0
6	4	6	5.00	20.900	1.0	4.0	2.0	4.0
7	6	8	5.00	20.900	1.0	4.0	2.0	4.0
8	8	10	5.00	20.900	1.0	4.0	2.0	4.0
9	10	12	5.00	20.900	1.0	4.0	2.0	4.0
10	12	14	5.00	20.900	1.0	4.0	2.0	4.0
11	14	16	5.00	20.900	1.0	4.0	2.0	4.0
12	16	18	5.00	20.900	1.0	4.0	2.0	4.0
13	18	20	5.00	20.900	1.0	4.0	2.0	4.0
14	20	22	5.00	20.900	1.0	4.0	2.0	4.0
15	3	5	5.89	14.400	1.0	4.0	2.0	4.0
16	5	7	5.89	14.400	1.0	4.0	2.0	4.0
17	7	9	5.89	14.400	1.0	4.0	2.0	4.0
18	9	11	5.89	14.400	1.0	4.0	2.0	4.0
19	11	13	5.89	14.400	1.0	4.0	2.0	4.0
20	13	15	5.89	14.400	1.0	4.0	2.0	4.0
21	15	17	5.89	14.400	1.0	4.0	2.0	4.0
22	17	19	5.89	14.400	1.0	4.0	2.0	4.0
23	19	21	5.89	14.400	1.0	4.0	2.0	4.0
24	21	23	5.89	14.400	1.0	4.0	2.0	4.0
25	4	5	2.38	2.170	1.0	4.0	2.0	4.0
26	6	7	1.88	.696	1.0	4.0	2.0	4.0
27	8	9	1.50	.591	1.0	4.0	2.0	4.0
28	10	11	1.50	.591	1.0	4.0	2.0	4.0
29	12	13	1.50	.591	1.0	4.0	2.0	4.0
30	14	15	1.50	.591	1.0	4.0	2.0	4.0
31	16	17	1.50	.591	1.0	4.0	2.0	4.0
32	18	19	1.88	.696	1.0	4.0	2.0	4.0
33	20	21	2.38	2.170	1.0	4.0	2.0	4.0
34	3	4	3.13	3.830	1.0	4.0	2.0	4.0
35	5	6	2.13	1.310	1.0	4.0	2.0	4.0
36	7	8	1.38	.277	1.0	4.0	2.0	4.0
37	9	10	1.38	.277	1.0	4.0	2.0	4.0
38	11	12	1.38	.277	1.0	4.0	2.0	4.0
39	12	15	1.38	.277	1.0	4.0	2.0	4.0

40	14	17	1.38	.277	1.0	4.0	2.0	4.0
41	16	19	1.38	.277	1.0	4.0	2.0	4.0
42	18	21	2.13	1.310	1.0	4.0	2.0	4.0
43	20	23	3.13	3.830	1.0	4.0	2.0	4.0

CONTROL INFORMATION

NUMBER OF JOINTS	= 24
NUMBER OF SUPPORT JOINTS	= 2
NUMBER OF MEMBERS	= 43
NUMBER OF INDEPENDENT LOAD CASES	= 4
NUMBER OF DEPENDENT LOAD CASES	= 5
TOTAL NUMBER OF SYSTEM DOF	= 66
STORAGE AVAILABLE	= 10000
STORAGE REQUIRED	= 8437
MATRIX SEMI-BANDWIDTH	= 12

APPLIED JOINT LOADS

JOINT	LOADING	X	Y	ZZ
1				
	1	.0	.0	.0
	2	.0	.0	.0
	3	.0	.0	.0
	4	.0	.0	.0
2				
	1	.0	-0.7	.0
	2	.0	.0	.0
	3	.0	.0	.0
	4	.0	.0	.0
3				
	1	.0	-0.8	.0
	2	.0	.0	.0
	3	.0	-2.4	.0
	4	-0.1	0.7	.0

4				
	1	.0	-.2	.0
	2	.0	.0	.0
	3	.0	.0	.0
	4	.0	.0	.0
5				
	1	.0	-1.5	.0
	2	.0	.0	.0
	3	.0	-4.8	.0
	4	-.1	1.3	.0
6				
	1	.0	-.2	.0
	2	.0	-6.4	.0
	3	.0	.0	.0
	4	.0	.0	.0
7				
	1	.0	-1.5	.0
	2	.0	.0	.0
	3	.0	-4.8	.0
	4	-.1	1.3	.0
8				
	1	.0	-.2	.0
	2	.0	.0	.0
	3	.0	.0	.0
	4	.0	.0	.0
9				
	1	.0	-1.5	.0
	2	.0	.0	.0
	3	.0	-4.8	.0
	4	-.1	1.3	.0
10				
	1	.0	-.2	.0
	2	.0	.0	.0
	3	.0	.0	.0
	4	.0	.0	.0
11				
	1	.0	-1.5	.0
	2	.0	.0	.0
	3	.0	-4.8	.0
	4	-.1	1.3	.0
12				
	1	0	-.2	.0
	2	0	-12.8	.0

265

	3	0	.0	.0
	4	0	.0	.0
13				
	1	0	-1.9	.0
	2	0	.0	.0
	3	0	-4.8	.0
	4	0	1.3	.0
14				
	1	0	-.2	.0
	2	0	.0	.0
	3	0	.0	.0
	4	0	.0	.0
15				
	1	0	-1.5	.0
	2	0	.0	.0
	3	0	-4.8	.0
	4	1	1.3	.0
16				
	1	0	-.2	.0
	2	0	.0	.0
	3	0	.0	.0
	4	0	.0	.0
17				
	1	0	-1.5	.0
	2	0	.0	.0
	3	0	-4.8	.0
	4	1	1.3	.0
18				
	1	0	-.2	.0
	2	0	-6.4	.0
	3	0	.0	.0
	4	0	.0	.0
19				
	1	0	-1.5	.0
	2	0	.0	.0
	3	0	-4.8	.0
	4	1	1.3	.0
20				
	1	0	-.2	.0
	2	0	.0	.0
	3	0	.0	.0
	4	0	.0	.0

21			
	1	0	-1.5	.0
	2	0	.0	.0
	3	0	-4.8	.0
	4	1	1.3	.0
22			
	1	.0	-.7	.0
	2	.0	.0	.0
	3	.0	.0	.0
	4	.0	.0	.0
23			
	1	.0	-.8	.0
	2	.0	.0	.0
	3	.0	-2.4	.0
	4	.1	.7	.0
24			
	1	.0	.0	.0
	2	.0	.0	.0
	3	.0	.0	.0
	4	.0	.0	.0

MEMBER FIXED ENDED FORCES

MEMBER NUMBER	LOADING NUMBER	JOINT NUMBER	X	Y	ZZ
1					
	1	1	.0	.0	.0
		2	.0	.0	.0
	2	1	.0	.0	.0
		2	.0	.0	.0
	3	1	.0	.0	.0
		2	.0	.0	.0
	4	1	.0	2.5	105.8
		2	.0	2.5	-105.8
2					
	1	2	.0	.0	.0
		3	.0	.0	.0
	2	2	.0	.0	.0
		3	.0	.0	.0

	3	2	.0	.0	.0
		3	.0	.0	.0
	4	2	.0	.5	4.9
		3	.0	.5	-4.9
3					
	1	24	.0	.0	.0
		22	.0	.0	.0
	2	24	.0	.0	.0
		22	.0	.0	.0
	3	24	.0	.0	.0
		22	.0	.0	.0
	4	24	.0	1.6	66.2
		22	.0	1.6	-66.2
4					
	1	22	.0	.0	.0
		23	.0	.0	.0
	2	22	.0	.0	.0
		23	.0	.0	.0
	3	22	.0	.0	.0
		23	.0	.0	.0
	4	22	.0	.3	3.0
		23	.0	.3	-3.0

NOTE by author:
All member fixed ended forces are zero for Members 5 to 43 and are not shown to save space, but the computer program prints all of these zero fixed ended forces.

SYSTEM DISPLACEMENTS

NOTE: All **Support Displacements** are shown as zero.
If any **Prescribed Support Displacement** was not zero
for any loading, the user will have to pencil in the
non-zero **Prescribed Support Displacements**.

DISPLACEMENTS

JOINT	LOADING	X	Y	ZZ
1				
	1	.0000	.0000	.0000
	2	.0000	.0000	.0000
	3	.0000	.0000	.0000
	4	.0000	.0000	.0000
	5	.0000	.0000	.0000
	6	.0000	.0000	.0000
	7	.0000	.0000	.0000
	8	.0000	.0000	.0000
	9	.0000	.0000	.0000
2				
	1	-.0286	-.0097	-.0006
	2	-.0517	-.0135	-.0011
	3	-.0799	-.0253	-.0017
	4	1.6000	.0078	-.0010
	5	-.0400	-.0135	-.0009
	6	-.1570	-.0458	-.0034
	7	-.1881	-.0588	-.0040
	8	1.9799	-.0208	-.0035
	9	2.0543	.0014	-.0019
3				
	1	.0128	-.0115	-.0009
	2	.0238	-.0163	-.0016
	3	.0358	-.0305	-.0024
	4	1.5771	.0093	.0011
	5	.0179	-.0161	-.0012
	6	.0713	-.0551	-.0048
	7	.0846	-.0708	-.0057
	8	2.0954	-.0252	-.0016
	9	2.0618	.0017	.0007
4				
	1	-.0289	-.0852	-.0011
	2	-.0522	-.1480	-.0021

3	-.0807	-.2370	-.0031
4	1.6022	.0585	.0011
5	-.0404	-.1193	-.0016
6	-.1585	-.4576	-.0062
7	-.1898	-.5554	-.0074
8	1.9818	-.2187	-.0025
9	2.0569	-.0006	.0004

5

1	.0158	-.0910	-.0011
2	.0295	-.1579	-.0020
3	.0442	-.2537	-.0030
4	1.5737	.0644	.0006
5	.0221	-.1274	-.0015
6	.0883	-.4887	-.0059
7	.1045	-.5941	-.0071
8	2.1017	-.2313	-.0030
9	2.0601	.0018	-.0001

6

1	.0247	-.1575	-.0009
2	.0453	-.2862	-.0016
3	.0690	-.4394	-.0025
4	.6005	.1142	.0006
5	.0346	-.2205	-.0013
6	.1366	-.8665	-.0049
7	.1627	-1.0351	-.0059
8	.9939	-.4033	-.0023
9	.0585	.0067	.0000

7

1	.0157	-.1631	-.0009
2	.0292	-.2912	-.0016
3	.0439	-.4555	-.0025
4	1.5729	.1197	.0008
5	.0220	-.2283	-.0012
6	.0876	-.8893	-.0048
7	.1037	-1.0701	-.0058
8	2.1001	-.4134	-.0021
9	2.0589	.0089	.0002

8

1	-.0177	-.2147	-.0006
2	-.0322	-.3790	-.0010
3	-.0493	-.5988	-.0018
4	1.5960	.1611	.0006
5	-.0248	-.3006	-.0009

	6	-.0974	-1.1634	-.0033
	7	-.1162	-1.4052	-.0041
	8	2.0129	-.5371	-.0014
	9	2.0589	.0162	.0002
9				
	1	.0129	-.2192	-.0006
	2	.0239	-.3854	-.0011
	3	.0361	-.6118	-.0017
	4	1.5742	.1662	.0005
	5	.0181	-.3069	-.0008
	6	.0718	-1.1855	-.0033
	7	.0853	-1.4345	-.0040
	8	2.0920	-.5455	-.0014
	9	2.0581	.0188	.0001
10				
	1	-.0091	-.2470	-.0002
	2	-.0170	-.4439	-.0008
	3	-.0254	-.6877	-.0007
	4	1.5899	.1901	.0002
	5	-.0128	-.3458	-.0003
	6	-.0508	-1.3504	-.0019
	7	-.0600	-1.6187	-.0017
	8	2.0347	-.6151	-.0007
	9	2.0587	.0248	.0001
11				
	1	.0075	-.2488	-.0002
	2	.0145	-.4498	-.0005
	3	.0209	-.6931	-.0004
	4	1.5779	.1930	.0002
	5	.0105	-.3483	-.0002
	6	.0427	-1.3648	-.0013
	7	.0497	-1.6324	-.0012
	8	2.0779	-.6191	-.0005
	9	2.0580	.0270	.0001
12				
	1	.0000	-.2538	.0000
	2	.0000	-.4789	.0000
	3	.0000	-.7050	.0000
	4	1.5830	.1994	.0001
	5	.0000	-.3553	.0000
	6	-.0001	-1.4233	.0000
	7	-.0001	-1.6721	.0000
	8	2.0578	-.6372	.0001

	9	2.0578	.0309	.0001
13				
	1	.0000	-.2518	.0000
	2	.0000	-.4670	.0000
	3	.0000	-.6988	.0000
	4	1.5836	.1976	.0000
	5	.0000	-.3525	.0000
	6	-.0001	-1.3988	.0000
	7	-.0001	-1.6537	.0000
	8	2.0586	-.6282	.0000
	9	2.0587	.0302	.0000
14				
	1	.0091	-.2470	.0002
	2	.0169	-.4439	.0008
	3	.0253	-.6877	.0007
	4	1.5753	.1993	-.0001
	5	.0128	-.3458	.0003
	6	.0507	-1.3504	.0019
	7	.0599	-1.6187	.0017
	8	2.0799	-.6030	.0008
	9	2.0561	.0369	.0001
15				
	1	-.0075	-.2488	.0002
	2	-.0145	-.4498	.0005
	3	-.0210	-.6931	.0004
	4	1.5898	.1992	-.0001
	5	-.0105	-.3483	.0002
	6	-.0428	-1.3648	.0013
	7	-.0498	-1.6324	.0012
	8	2.0400	-.6110	.0006
	9	2.0600	.0351	.0001
16				
	1	.0177	-.2147	.0006
	2	.0322	-.3790	.0010
	3	.0492	-.5988	.0018
	4	1.5675	.1785	-.0004
	5	.0247	-.3006	.0009
	6	.0973	-1.1634	.0033
	7	.1161	-1.4052	.0041
	8	2.0997	-.5145	.0016
	9	2.0537	.0388	.0000

17			
1	-.0130	-.2192	.0006
2	-.0239	-.3854	.0011
3	-.0362	-.6118	.0017
4	1.5948	.1804	-.0004
5	-.0182	-.3069	.0008
6	-.0719	-1.1855	.0033
7	-.0854	-1.4345	.0040
8	2.0277	-.5271	.0015
9	2.0616	.0373	.0000
18			
1	.0247	-.1575	.0009
2	.0453	-.2862	.0016
3	.0689	-.4394	.0025
4	1.5605	.1356	-.0008
5	.0346	-.2205	.0013
6	.1365	-.8665	.0049
7	.1626	-1.0351	.0059
8	2.1153	-.3755	.0021
9	2.0508	.0345	-.0002
19			
1	-.0157	-.1631	.0009
2	-.0293	-.2912	.0016
3	-.0440	-.4555	.0025
4	1.5979	.1389	-.0006
5	-.0220	-.2283	.0012
6	-.0877	-.8893	.0048
7	-.1039	-1.0701	.0058
8	2.0218	-.3885	.0023
9	2.0631	.0338	.0000
20			
1	.0288	-.0852	.0011
2	.0522	-.1480	.0021
3	.0806	-.2370	.0031
4	1.5549	.0766	-.0006
5	.0404	-.1193	.0016
6	.1584	-.4576	.0062
7	.1897	-.5554	.0074
8	2.1224	-.1952	.0031
9	2.0474	.0229	.0002

273

21				
	1	-.0158	-.0910	.0011
	2	-.0296	-.1579	.0020
	3	-.0443	-.2537	.0030
	4	1.5989	.0799	-.0011
	5	-.0222	-.1274	.0015
	6	-.0885	-.4887	.0059
	7	-.1047	-.5941	.0071
	8	2.0227	-.2112	.0024
	9	2.0644	.0219	-.0005
22				
	1	.0286	-.0097	.0006
	2	.0516	-.0135	.0011
	3	.0799	-.0253	.0017
	4	1.5523	.0061	-.0022
	5	.0400	-.0135	.0009
	6	.1569	-.0458	.0034
	7	.1879	-.0588	.0040
	8	2.1181	-.0230	-.0007
	9	2.0438	-.0008	-.0023
23				
	1	-.0128	-.0115	.0009
	2	-.0238	-.0163	.0016
	3	-.0359	-.0305	.0024
	4	1.5975	.0076	-.0002
	5	-.0180	-.0161	.0012
	6	-.0714	-.0551	.0048
	7	-.0847	-.0708	.0057
	8	2.0315	-.0274	.0027
	9	2.0652	-.0005	.0005
24				
	1	.0000	.0000	.0000
	2	.0000	.0000	.0000
	3	.0000	.0000	.0000
	4	.0000	.0000	.0000
	5	.0000	.0000	.0000
	6	.0000	.0000	.0000
	7	.0000	.0000	.0000
	8	.0000	.0000	.0000
	9	.0000	.0000	.0000

REACTIONS

JOINT	LOADING	X	Y	ZZ
1				
	1	.2	9.2	-21.5
	2	.4	12.8	-38.9
	3	.6	24.0	-60.1
	4	-5.7	-7.4	512.4
	5	.3	12.8	-30.1
	6	1.2	43.5	-118.0
	7	1.5	55.8	-141.3
	8	-6.6	19.8	590.8
	9	-7.1	-1.4	646.7
24				
	1	-.2	9.2	21.5
	2	-.4	12.8	38.8
	3	-.6	24.0	60.1
	4	-4.3	-5.8	433.9
	5	-.3	12.8	30.1
	6	-1.2	43.5	117.9
	7	-1.5	55.8	141.3
	8	-6.4	21.9	639.3
	9	-5.8	.7	583.4

MEMBER END FORCES

MEMBER NUMBER	LOADING NUMBER	JOINT NUMBER	X	Y	ZZ
1					
	1	1	9.2	-.2	-21.5
		2	-9.2	.2	-35.3
	2	1	12.8	-.4	-38.9
		2	-12.8	.4	-63.8
	3	1	24.0	-.6	-60.1
		2	-24.0	.6	-98.7
	4	1	-7.4	5.7	512.4
		2	7.4	-.6	277.7
	5	1	12.8	-.3	-30.1
		2	-12.8	.3	-49.4
	6	1	43.5	-1.2	-118.0
		2	-43.5	1.2	-193.8
	7	1	55.8	-1.5	-141.3
		2	-55.8	1.5	-232.1
	8	1	19.8	6.6	590.8
		2	-19.8	.0	237.4
	9	1	-1.4	7.1	646.7
		2	1.4	-.6	329.2
2					
	1	2	8.2	.3	21.7
		3	-8.2	-.3	-5.5
	2	2	12.4	.6	41.6
		3	-12.4	-.6	-8.2
	3	2	23.3	.8	60.5
		3	-23.3	-.8	-15.6
	4	2	-6.5	-3.8	-226.3
		3	6.5	4.9	-10.5
	5	2	11.5	.4	30.4
		3	-11.5	-.4	-7.7
	6	2	41.3	1.8	122.9
		3	-41.3	-1.8	-27.6
	7	2	53.3	2.0	143.7
		3	-53.3	-2.0	-35.6
	8	2	19.3	-3.9	-217.0
		3	-19.3	5.3	-32.1
	9	2	-1.0	-4.7	-274.6
		3	1.0	6.1	-18.5

1	24	9.2	.2	21.5
	22	-9.2	-.2	35.3
2	24	12.8	.4	38.8
	22	-12.8	-.4	63.8
3	24	24.0	.6	60.1
	22	-24.0	-.6	98.7
4	24	-5.8	4.3	433.9
	22	5.8	1.1	252.3
5	24	12.8	.3	30.1
	22	-12.8	-.3	49.4
6	24	43.5	1.2	117.9
	22	-43.5	1.2	193.8
7	24	55.8	1.5	141.3
	22	-55.8	1.5	232.1
8	24	21.9	6.4	639.3
	22	-21.9	2.3	451.6
9	24	.7	5.8	583.4
	22	-.7	1.7	359.8

1	22	8.2	-.3	-21.7
	23	-8.2	.3	5.5
2	22	12.4	-.6	-41.6
	23	12.4	.6	8.2
3	22	23.3	-.8	-60.5
	23	23.3	.8	15.6
4	22	-6.4	-4.1	-217.6
	23	6.4	4.7	-20.2
5	22	11.5	-.4	-30.4
	23	11.5	.4	7.7
6	22	41.3	-1.8	-122.9
	23	41.3	1.8	27.6
7	22	53.3	-2.0	-143.7
	23	53.3	2.0	35.6
8	22	19.4	-6.4	-360.0
	23	19.4	7.2	-7.8
9	22	-.9	-5.6	-302.4
	23	.9	6.4	-21.3

1	2	.5	.3	13.6
	4	-.5	-.3	5.1
2	2	1.0	.4	22.2
	4	-1.0	-.4	5.9

3	2	1.5	.7	38.1
	4	-1.5	-.7	14.2
4	2	-4.5	-.9	-51.4
	4	4.5	.9	-15.6
5	2	.7	.4	19.0
	4	-.7	-.4	7.1
6	2	3.0	1.3	70.9
	4	-3.0	-1.3	22.7
7	2	3.5	1.7	88.4
	4	-3.5	-1.7	31.7
8	2	-3.9	-.3	-20.4
	4	3.9	.3	-4.2
9	2	-5.3	-1.0	-54.7
	4	5.3	1.0	-15.8

6

1	4	-8.3	.0	-2.0
	6	8.3	.0	1.7
2	4	-13.9	.1	.0
	6	13.9	-.1	7.8
3	4	-23.5	.0	-5.7
	6	23.5	.0	4.9
4	4	3.3	.1	8.9
	6	-3.3	-.1	1.2
5	4	-11.7	.0	-2.8
	6	11.7	.0	2.4
6	4	-44.0	.2	-5.3
	6	44.0	-.2	17.0
7	4	-54.5	.0	11.5
	6	54.5	.0	13.8
8	4	-24.4	.2	6.3
	6	24.4	-.2	9.9
9	4	-3.2	.2	9.8
	6	3.2	-.2	3.1

7

1	6	-14.1	.0	-.9
	8	14.1	.0	3.4
2	6	-26.4	.0	-6.6
	8	26.4	.0	3.0
3	6	-39.7	.1	-2.5
	8	39.7	-.1	9.7
4	6	9.1	-.1	-1.8
	8	-9.1	.1	-3.1
5	6	-19.8	.1	-1.2

		8	19.8	-.1	4.8
	6	6	-79.0	.0	-12.8
		8	79.0	.0	13.8
	7	6	-93.7	.2	-8.4
		8	93.7	-.2	21.1
	8	6	-38.2	.0	-7.9
		8	38.2	.0	6.4
	9	6	-.9	-.1	-3.1
		8	.9	.1	-.9
8					
	1	8	-17.2	.0	-3.1
		10	17.2	.0	3.5
	2	8	-30.7	.0	-2.5
		10	30.7	.0	1.9
	3	8	-48.2	.0	-8.7
		10	48.2	.0	10.1
	4	8	12.4	.0	2.7
		10	-12.4	.0	-2.8
	5	8	-24.1	.0	-4.3
		10	24.1	.0	5.0
	6	8	-93.8	.0	-12.1
		10	93.8	.0	12.5
	7	8	-113.1	.0	-18.9
		10	113.1	.0	21.5
	8	8	-44.0	.0	-5.8
		10	44.0	.0	6.7
	9	8	.6	.0	.7
		10	-.6	.0	-.4
9					
	1	10	-18.4	.0	-3.5
		12	18.4	.0	.6
	2	10	-34.1	.1	-1.5
		12	34.1	-.1	11.5
	3	10	-51.0	-.1	-10.1
		12	51.0	.1	1.0
	4	10	14.0	.0	2.6
		12	-14.0	.0	-.3
	5	10	-25.7	-.1	-4.9
		12	25.7	.1	.9
	6	10	-102.2	.1	-11.7
		12	102.2	-.1	19.7
	7	10	-120.8	-.2	-21.2
		12	120.8	.2	8.1

8	10	-46.5	.0	-6.6
	12	46.5	.0	6.7
9	10	1.6	.0	.3
	12	-1.6	.0	.2
10				
1	12	-18.4	.0	-.6
	14	18.4	.0	3.5
2	12	-34.1	-.1	-11.5
	14	34.1	.1	1.5
3	12	-51.0	.1	-1.0
	14	51.0	-.1	10.1
4	12	15.5	.0	.2
	14	-15.5	.0	-3.1
5	12	-25.7	.1	-.9
	14	25.7	-.1	4.9
6	12	-102.2	-.1	-19.7
	14	102.2	.1	11.7
7	12	-120.8	.2	-8.1
	14	120.8	-.2	21.2
8	12	-44.5	.0	-6.8
	14	44.5	.0	6.0
9	12	3.6	.0	-.4
	14	-3.6	.0	-.9
11				
1	14	-17.2	.0	-3.5
	16	17.2	.0	3.1
2	14	-30.7	.0	-1.9
	16	30.7	.0	2.5
3	14	-48.2	.0	-10.1
	16	48.2	.0	8.7
4	14	15.6	.0	3.0
	16	-15.6	.0	-2.2
5	14	-24.1	.0	-5.0
	16	24.1	.0	4.3
6	14	-93.8	.0	-12.4
	16	93.8	.0	12.1
7	14	-113.1	.0	-21.5
	16	113.1	.0	18.9
8	14	-39.9	.0	-6.4
	16	39.9	.0	6.5
9	14	4.7	.0	.7
	16	-4.7	.0	-.1

12					
	1	16	-14.1	.0	-3.4
		18	14.1	.0	.9
	2	16	-26.4	.0	-3.0
		18	26.4	.0	6.6
	3	16	-39.7	-.1	-9.7
		18	39.7	.1	2.5
	4	16	14.3	.0	2.3
		18	-14.3	.0	-3.5
	5	16	-19.8	-.1	-4.8
		18	19.8	.1	1.2
	6	16	-79.0	.0	13.8
		18	79.0	.0	12.8
	7	16	-93.7	-.2	21.1
		18	93.7	.2	8.4
	8	16	-31.5	-.1	-7.5
		18	31.5	.1	1.0
	9	16	5.8	-.1	-.2
		18	-5.8	.1	-3.8
13					
	1	18	-8.3	.0	-1.7
		20	8.3	.0	2.0
	2	18	-13.9	-.1	-7.8
		20	13.9	.1	.0
	3	18	-23.5	.0	-4.9
		20	23.5	.0	5.7
	4	18	11.2	.1	4.2
		20	-11.2	-.1	6.5
	5	18	-11.7	.0	-2.4
		20	11.7	.0	2.8
	6	18	-44.0	-.2	17.0
		20	44.0	.2	5.3
	7	18	-54.5	.0	13.8
		20	54.5	.0	11.5
	8	18	-14.2	.2	-2.9
		20	14.2	-.2	13.7
	9	18	7.0	.2	4.0
		20	-7.0	-.2	10.3
14					
	1	20	.5	-.3	-5.1
		22	-.5	.3	13.6
	2	20	1.0	-.4	-5.9
		22	-1.0	.4	22.2

281

	3	20	1.5	-.7	14.2
		22	-1.5	.7	38.1
	4	20	5.2	-.6	-9.0
		22	-5.2	.6	34.8
	5	20	.7	-.4	-7.1
		22	-.7	.4	19.0
	6	20	3.0	-1.3	22.7
		22	-3.0	1.3	70.9
	7	20	3.5	-1.7	31.7
		22	-3.5	1.7	88.4
	8	20	8.6	-1.7	27.8
		22	-8.6	1.7	91.7
	9	20	7.2	-1.0	16.2
		22	-7.2	1.0	57.4
15					
	1	3	8.5	.1	5.6
		5	-8.5	-.1	3.3
	2	3	14.2	.2	8.7
		5	-14.2	-.2	4.4
	3	3	24.0	.4	16.0
		5	-24.0	-.4	9.3
	4	3	-2.8	.1	6.7
		5	2.8	-.1	1.1
	5	3	11.9	.2	7.9
		5	-11.9	-.2	4.6
	6	3	45.0	.6	28.7
		5	-45.0	-.6	15.6
	7	3	55.8	.8	36.7
		5	-55.8	-.8	21.0
	8	3	25.7	.6	27.8
		5	-25.7	-.6	12.2
	9	3	4.0	.3	13.8
		5	-4.0	-.3	4.3
16					
	1	5	14.4	.0	-.6
		7	-14.4	.0	1.5
	2	5	26.9	.1	.3
		7	-26.9	-.1	4.9
	3	5	40.4	.0	-1.9
		7	-40.4	.0	4.2
	4	5	-8.8	-.1	-3.2
		7	8.8	.1	-1.8
	5	5	20.2	.0	-.9

		7	-20.2	.0	2.1
	6	5	80.5	.1	-1.2
		7	-80.5	-.1	11.7
	7	5	95.4	.1	-3.6
		7	-95.4	-.1	10.9
	8	5	39.5	.0	-5.7
		7	-39.5	.0	4.0
	9	5	1.5	-.1	-4.7
		7	-1.5	.1	-1.0
17					
	1	7	17.5	.0	-.9
		9	-17.5	.0	2.4
	2	7	31.2	.0	-3.9
		9	-31.2	.0	1.6
	3	7	49.0	.1	-2.5
		9	-49.0	-.1	6.8
	4	7	-12.2	.0	1.1
		9	12.2	.0	-1.6
	5	7	24.5	.0	-1.2
		9	-24.5	.0	3.4
	6	7	95.4	.0	-8.6
		9	-95.4	.0	8.8
	7	7	115.0	.1	-7.0
		9	-115.0	-.1	14.5
	8	7	45.2	.0	-2.8
		9	-45.2	.0	5.0
	9	7	-0.1	.0	.7
		9	0.1	.0	.1
18					
	1	9	18.7	.0	-2.1
		11	-18.7	.0	2.9
	2	9	34.7	.1	-1.0
		11	-34.7	-.1	5.3
	3	9	51.8	.0	-5.8
		11	-51.8	.0	8.4
	4	9	-13.9	.0	1.2
		11	13.9	.0	-2.7
	5	9	26.1	.0	-2.9
		11	-26.1	.0	4.1
	6	9	103.8	.1	-7.0
		11	-103.8	-.1	16.1
	7	9	122.7	.1	-12.3
		11	-122.7	-.1	19.6

283

8	9	47.5	.0	-4.3
	11	-47.5	.0	6.9
9	9	-1.3	.0	-.3
	11	1.3	.0	-.8

19

1	11	18.3	-.1	-2.9
	13	-18.3	.1	-.9
2	11	37.6	.0	-5.0
	13	-37.6	.0	1.3
3	11	50.4	-.2	-8.5
	13	-50.4	.2	-3.3
4	11	-14.4	.0	2.6
	13	14.4	.0	1.0
5	11	25.6	-.1	-4.1
	13	-25.6	.1	-1.2
6	11	107.3	-.2	-15.7
	13	-107.3	.2	-.6
7	11	121.5	-.3	-19.6
	13	-121.5	.3	-5.7
8	11	47.3	-.1	-6.9
	13	-47.3	.1	-.8
9	11	-2.2	.0	.7
	13	2.2	.0	.4

20

1	13	18.3	.1	.9
	15	-18.3	-.1	2.9
2	13	37.6	.0	-1.3
	15	-37.6	.0	5.0
3	13	50.4	.2	3.3
	15	-50.4	-.2	8.5
4	13	-14.4	.0	-1.0
	15	14.4	.0	-2.3
5	13	25.6	.1	1.2
	15	-25.6	-.1	4.1
6	13	107.3	.2	.6
	15	-107.3	-.2	15.7
7	13	121.5	.3	5.7
	15	-121.5	-.3	19.6
8	13	47.3	.1	.7
	15	-47.3	-.1	7.3
9	13	-2.2	.0	-.5
	15	2.2	.0	-.3

21					
	1	15	18.7	.0	-2.9
		17	-18.7	.0	2.1
	2	15	34.7	.1	-5.3
		17	-34.7	.1	1.0
	3	15	51.8	.0	-8.4
		17	-51.8	.0	5.8
	4	15	-15.4	.0	2.1
		17	15.4	.0	-2.1
	5	15	26.1	.0	-4.1
		17	-26.1	.0	2.9
	6	15	103.8	-.1	-16.1
		17	-103.8	.1	7.0
	7	15	122.7	-.1	-19.6
		17	-122.7	.1	12.3
	8	15	45.6	-.1	-7.6
		17	-45.6	.1	3.2
	9	15	-3.3	.0	.2
		17	3.3	.0	-.8
22					
	1	17	17.5	.0	-2.4
		19	-17.5	.0	.9
	2	17	31.2	.0	-1.6
		19	-31.2	.0	3.9
	3	17	49.0	-.1	-6.8
		19	-49.0	.1	2.5
	4	17	-15.4	.0	2.2
		19	15.4	.0	-.3
	5	17	24.5	.0	-3.4
		19	-24.5	.0	1.2
	6	17	95.4	.0	-8.8
		19	-95.4	.0	8.6
	7	17	115.0	-.1	-14.5
		19	-115.0	.1	7.0
	8	17	41.0	.0	-4.2
		19	-41.0	.0	3.9
	9	17	-4.3	.0	.7
		19	4.3	.0	.4
23					
	1	19	14.4	.0	-1.5
		21	-14.4	.0	.6
	2	19	26.9	-.1	-4.9
		21	-26.9	.1	-.3

	3	19	40.4	.0	-4.2
		21	-40.4	.0	1.9
	4	19	-14.0	-.1	.5
		21	14.0	.1	-4.6
	5	19	20.2	.0	-2.1
		21	-20.2	.0	.9
	6	19	80.5	-.1	-11.7
		21	-80.5	.1	1.2
	7	19	95.4	-.1	-10.9
		21	-95.4	.1	3.6
	8	19	32.7	-.1	-5.6
		21	-32.7	.1	-4.4
	9	19	-5.2	-.1	-.6
		21	5.2	.1	-5.4
24					
	1	21	8.5	-.1	-3.3
		23	-8.5	.1	-5.6
	2	21	14.2	-.2	-4.4
		23	-14.2	.2	-8.7
	3	21	24.0	-.4	-9.3
		23	-24.0	.4	-16.0
	4	21	-10.8	.3	6.6
		23	10.8	-.3	16.3
	5	21	11.9	-.2	-4.6
		23	-11.9	.2	-7.9
	6	21	45.0	-.6	-15.6
		23	-45.0	.6	-28.7
	7	21	55.8	-.8	-21.0
		23	-55.8	.8	-36.7
	8	21	15.3	.0	-2.2
		23	-15.3	.0	2.1
	9	21	-6.3	.3	5.6
		23	6.3	-.3	16.2
25					
	1	4	6.7	-.1	-2.3
		5	-6.7	.1	-2.2
	2	4	11.3	-.1	-4.2
		5	-11.3	.1	-4.0
	3	4	19.3	-.2	-6.4
		5	-19.3	.2	-6.1
	4	4	-6.8	.1	3.0
		5	6.8	-.1	2.0
	5	4	9.4	-.1	-3.2

		5	-9.4	.1	-3.0
	6	4	35.8	-.4	-12.7
		5	-35.8	.4	-12.1
	7	4	44.5	-.5	-15.0
		5	-44.5	.5	-14.4
	8	4	14.6	-.2	-4.1
		5	-14.6	.2	-5.1
	9	4	-2.8	.0	1.8
		5	2.8	.0	.6
26					
	1	6	4.6	.0	-.5
		7	-4.6	.0	-.5
	2	6	4.1	.0	-.9
		7	-4.1	.0	-.8
	3	6	13.3	.0	-1.4
		7	-13.3	.0	-1.4
	4	6	-4.6	.0	.5
		7	4.6	.0	.5
	5	6	6.5	.0	-.7
		7	-6.5	.0	-.7
	6	6	18.8	-.1	-2.7
		7	-18.8	.1	-2.7
	7	6	28.9	-.1	-3.4
		7	-28.9	.1	-3.3
	8	6	8.3	.0	-1.2
		7	-8.3	.0	-1.0
	9	6	-1.8	.0	.2
		7	1.8	.0	.2
27					
	1	8	2.7	.0	-.3
		9	-2.7	.0	-.3
	2	8	3.9	.0	-.4
		9	-3.9	.0	-.4
	3	8	7.9	.0	-.8
		9	-7.9	.0	-.8
	4	8	-3.1	.0	.3
		9	3.1	.0	.3
	5	8	3.8	.0	-.4
		9	-3.8	.0	-.4
	6	8	13.4	.0	-1.4
		9	-13.4	.0	-1.4
	7	8	17.8	.0	-1.8
		9	-17.8	.0	-1.7

8	8	5.1	.0	-.5
	9	-5.1	.0	-.5
9	8	-1.6	.0	.2
	9	1.6	.0	.2

28

1	10	1.0	.0	.0
	11	-1.0	.0	.0
2	10	3.3	.0	-.4
	11	-3.3	.0	-.3
3	10	3.0	.0	.0
	11	-3.0	.0	.1
4	10	-1.6	.0	.1
	11	1.6	.0	.1
5	10	1.4	.0	.0
	11	-1.4	.0	.0
6	10	8.0	.0	-.6
	11	-8.0	.0	-.4
7	10	7.7	.0	-.2
	11	-7.7	.0	.0
8	10	2.2	.0	-.1
	11	-2.2	.0	.0
9	10	-1.2	.0	.1
	11	1.2	.0	.1

29

1	12	-1.0	.0	.0
	13	1.0	.0	.0
2	12	-6.1	.0	.0
	13	6.1	.0	.0
3	12	-3.3	.0	.0
	13	3.3	.0	.0
4	12	1.0	.0	.1
	13	-1.0	.0	.1
5	12	-1.4	.0	.0
	13	1.4	.0	.0
6	12	-12.7	.0	.0
	13	12.7	.0	.0
7	12	-9.5	.0	.0
	13	9.5	.0	.0
8	12	-4.7	.0	.1
	13	4.7	.0	.1
9	12	.3	.0	.1
	13	-.3	.0	.1

30					
	1	14	1.0	.0	.0
		15	-1.0	.0	.0
	2	14	3.3	.0	.4
		15	-3.3	.0	.3
	3	14	3.0	.0	.0
		15	-3.0	.0	-.1
	4	14	.1	.0	.1
		15	-.1	.0	.1
	5	14	1.4	.0	.0
		15	-1.4	.0	.0
	6	14	8.0	.0	.6
		15	-8.0	.0	.4
	7	14	7.7	.0	.2
		15	-7.7	.0	.0
	8	14	4.5	.0	.3
		15	-4.5	.0	.2
	9	14	1.0	.0	.1
		15	-1.0	.0	.1
31					
	1	16	2.7	.0	.3
		17	-2.7	.0	.3
	2	16	3.9	.0	.4
		17	-3.9	.0	.4
	3	16	7.9	.0	.8
		17	-7.9	.0	.8
	4	16	-1.2	.0	-.1
		17	1.2	.0	-.1
	5	16	3.8	.0	.4
		17	-3.8	.0	.4
	6	16	13.4	.0	1.4
		17	-13.4	.0	1.4
	7	16	17.8	.0	1.8
		17	-17.8	.0	1.7
	8	16	7.6	.0	.8
		17	-7.6	.0	.8
	9	16	.9	.0	.2
		17	-.9	.0	.1
32					
	1	18	4.6	.0	.5
		19	-4.6	.0	.5
	2	18	4.1	.0	.9
		19	-4.1	.0	.8

	3	18	13.3	.0	1.4
		19	-13.3	.0	1.4
	4	18	-2.7	.0	-.3
		19	2.7	.0	-.2
	5	18	6.5	.0	.7
		19	-6.5	.0	.7
	6	18	18.8	.1	2.7
		19	-18.8	.1	2.7
	7	18	28.9	.1	3.4
		19	-28.9	.1	3.3
	8	18	10.7	.0	1.4
		19	-10.7	.0	1.4
	9	18	.6	.0	.1
		19	-.6	.0	.2
33					
	1	20	6.7	.1	2.3
		21	-6.7	-.1	2.2
	2	20	11.3	.1	4.2
		21	-11.3	-.1	4.0
	3	20	19.3	.2	6.4
		21	-19.3	-.2	6.1
	4	20	-3.8	.0	-.4
		21	3.8	.0	-1.3
	5	20	9.4	.1	3.2
		21	-9.4	-.1	3.0
	6	20	35.8	.4	12.7
		21	-35.8	-.4	12.1
	7	20	44.5	.5	15.0
		21	-44.5	-.5	14.4
	8	20	18.4	.2	7.5
		21	-18.4	-.2	5.9
	9	20	1.1	.0	1.5
		21	-1.1	.0	.2
34					
	1	3	-11.0	.0	-.1
		4	11.0	.0	-.8
	2	3	-18.4	.0	-.5
		4	18.4	.0	-1.7
	3	3	-30.9	.0	-.4
		4	30.9	.0	-2.1
	4	3	9.6	.1	3.8
		4	-9.6	-.1	3.7
	5	3	-15.4	.0	-.2

		4	15.4	.0	-1.1
	6	3	-58.1	-.1	-1.2
		4	58.1	.1	-4.7
	7	3	-71.9	-.1	-1.1
		4	71.9	.1	-5.2
	8	3	-25.4	.1	4.3
		4	25.4	-.1	2.0
	9	3	2.6	.1	4.8
		4	-2.6	-.1	4.2
35					
	1	5	-7.5	.0	-.5
		6	7.5	.0	-.3
	2	5	-16.2	.0	-.7
		6	16.2	.0	-.4
	3	5	-21.0	.0	-1.3
		6	21.0	.0	-.9
	4	5	7.4	.0	.1
		6	-7.4	.0	.1
	5	5	-10.5	.0	-.7
		6	10.5	.0	-.5
	6	5	-45.5	.0	-2.3
		6	45.5	.0	-1.5
	7	5	-50.7	-.1	-3.0
		6	50.7	.1	-2.1
	8	5	-18.0	.0	-1.4
		6	18.0	.0	-.9
	9	5	2.9	.0	-.3
		6	-2.9	.0	-.1
36					
	1	7	-4.2	.0	-.1
		8	4.2	.0	-.1
	2	7	-5.8	.0	-.1
		8	5.8	.0	-.1
	3	7	-11.5	.0	-.3
		8	11.5	.0	-.1
	4	7	4.5	.0	.1
		8	-4.5	.0	.1
	5	7	-5.9	.0	-.1
		8	5.9	.0	-.1
	6	7	-20.0	.0	-.5
		8	20.0	.0	-.2
	7	7	-26.3	.0	-.6
		8	26.3	.0	-.3

8	7	-7.9	.0	-.2
	8	7.9	.0	-.1
9	7	2.0	.0	.1
	8	-2.0	.0	.1
37				
1	9	-1.6	.0	-.1
	10	1.6	.0	.0
2	9	-4.9	.0	-.1
	10	4.9	.0	-.1
3	9	-4.0	.0	-.2
	10	4.0	.0	.0
4	9	2.3	.0	.1
	10	-2.3	.0	.0
5	9	-2.2	.0	-.1
	10	2.2	.0	.0
6	9	-11.8	.0	-.4
	10	11.8	.0	-.2
7	9	-10.8	.0	-.4
	10	10.8	.0	-.1
8	9	-3.4	.0	-.2
	10	3.4	.0	.0
9	9	1.5	.0	.0
	10	-1.5	.0	.0
38				
1	11	.5	.0	.0
	12	-.5	.0	.0
2	11	-4.3	.0	.0
	12	4.3	.0	.0
3	11	2.0	.0	.0
	12	-2.0	.0	.0
4	11	.5	.0	.0
	12	-.5	.0	.0
5	11	.8	.0	.0
	12	-.8	.0	.0
6	11	-5.3	.0	-.1
	12	5.3	.0	.1
7	11	1.7	.0	-.1
	12	-1.7	.0	.1
8	11	.1	.0	.0
	12	-.1	.0	.1
9	11	1.1	.0	.0
	12	-1.1	.0	.0

39					
	1	12	.5	.0	.0
		15	-.5	.0	.0
	2	12	-4.3	.0	.0
		15	4.3	.0	.0
	3	12	2.0	.0	.0
		15	-2.0	.0	.0
	4	12	-1.7	.0	.0
		15	1.7	.0	.0
	5	12	.8	.0	.0
		15	-.8	.0	.0
	6	12	-5.3	.0	-.1
		15	5.3	.0	.1
	7	12	1.7	.0	-.1
		15	-1.7	.0	.1
	8	12	-2.7	.0	.0
		15	2.7	.0	.1
	9	12	-1.7	.0	.0
		15	1.7	.0	.0
40					
	1	14	-1.6	.0	.0
		17	1.6	.0	.1
	2	14	-4.9	.0	.1
		17	4.9	.0	.1
	3	14	-4.0	.0	.0
		17	4.0	.0	.2
	4	14	-.2	.0	.0
		17	.2	.0	.0
	5	14	-2.2	.0	.0
		17	2.2	.0	.1
	6	14	-11.8	.0	.2
		17	11.8	.0	.4
	7	14	-10.8	.0	.1
		17	10.8	.0	.4
	8	14	-6.6	.0	.1
		17	6.6	.0	.2
	9	14	-1.6	.0	.0
		17	1.6	.0	.0
41					
	1	16	-4.2	.0	.1
		19	4.2	.0	.1
	2	16	-5.8	.0	.1
		19	5.8	.0	.1

3	16	-11.5	.0	.1
	19	11.5	.0	.3
4	16	1.8	.0	.0
	19	-1.8	.0	.0
5	16	-5.9	.0	.1
	19	5.9	.0	.1
6	16	-20.0	.0	.2
	19	20.0	.0	.5
7	16	-26.3	.0	.3
	19	26.3	.0	.6
8	16	-11.4	.0	.2
	19	11.4	.0	.3
9	16	-1.5	.0	.0
	19	1.5	.0	.0

42

1	18	-7.5	.0	.3
	21	7.5	.0	.5
2	18	-16.2	.0	.4
	21	16.2	.0	.7
3	18	-21.0	.0	.9
	21	21.0	.0	1.3
4	18	4.0	.0	-.4
	21	-4.0	.0	-.7
5	18	-10.5	.0	.5
	21	10.5	.0	.7
6	18	-45.5	.0	1.5
	21	45.5	.0	2.3
7	18	-50.7	.1	2.1
	21	50.7	-.1	3.0
8	18	-22.4	.0	.5
	21	22.4	.0	.7
9	18	-1.5	.0	-.2
	21	1.5	.0	-.4

43

1	20	-11.0	.0	.8
	23	11.0	.0	.1
2	20	-18.4	.0	1.7
	23	18.4	.0	.5
3	20	-30.9	.0	2.1
	23	30.9	.0	.4
4	20	7.5	.1	2.9
	23	-7.5	-.1	3.8
5	20	-15.4	.0	1.1

	23	15.4	.0	.2
6	20	-58.1	.1	4.7
	23	58.1	-.1	1.2
7	20	-71.9	.1	5.2
	23	71.9	-.1	1.1
8	20	-28.2	.1	6.5
	23	28.2	-.1	5.6
9	20	-.2	.1	4.4
	23	.2	-.1	5.1

Centroidal Axes;
Moments and Product of Inertia;
Transfer Axes Formulas;
Principal Axes and Mohr's Circle;
Radii of Gyration;
Flexural Formula; Biaxial Bending.

B.1 NOTATION:

X and Y are orthogonal, reference axes.
x and y are **centroidal axes** parallel to the X and Y axes.

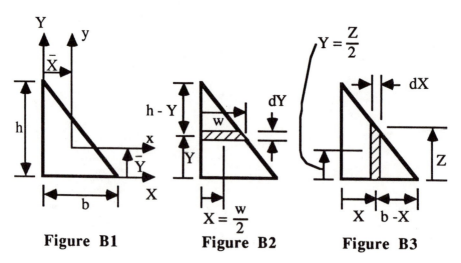

Figure B1   Figure B2   Figure B3

From similar triangles:

Figure 2: $\dfrac{w}{b} = \dfrac{h-Y}{h}$; $w = \dfrac{b}{h}(h-Y)$

Figure 3: $\dfrac{Z}{h} = \dfrac{b-X}{b}$; $Z = \dfrac{h}{b}(b-X)$

B.2 CENTROIDAL AXES

In Figure B1, the origin of the **centroidal axes**(x,y) is at $X = \bar{X}$ and $Y = \bar{Y}$ where $\bar{X}$ and $\bar{Y}$ are defined as follows:

$$A = \int dA = \int_0^h w \, dy = \frac{b}{h} \int_0^h (h - Y) \, dY = \frac{1}{2} \, bh$$

$$\bar{X} = \frac{\int X \, dA}{A} = \frac{\int_0^b XZ \, dX}{A} = \frac{\frac{h}{b} \int_0^b X(b - X) \, dX}{\frac{1}{2} \, bh} = \frac{1}{3} \, b$$

$$\bar{Y} = \frac{\int Y \, dA}{A} = \frac{\int_0^h Yw \, dY}{A} = \frac{\frac{b}{h} \int_0^h Y(h - Y) \, dY}{\frac{1}{2} \, bh} = \frac{1}{3} \, h$$

B.3 MOMENTS AND PRODUCT OF INERTIA

For the **reference axes(X,Y)**, the definition of the **moments of inertia** (I_X and I_Y) and the **product of inertia** (I_{XY}) are:
Using Figure B2:

$$I_X = \int Y^2 \, dA = \int_0^h Y^2 w \, dY = \frac{b}{h}\int_0^h Y^2(h-Y) \, dY = \frac{1}{12} bh^3$$

Using Figure B3:

$$I_Y = \int X^2 \, dA = \int_0^b X^2 Z \, dX = \frac{h}{b}\int_0^b X^2(b-X) \, dX = \frac{1}{12} hb^3$$

Using Figure B2:

$$I_{XY} = \int XY \, dA = \int_0^h XYw \, dY = \int_0^h \frac{1}{2}wYw \, dY$$

$$I_{XY} = \frac{1}{2}\frac{b^2}{h^2}\int_0^h Y(h-Y)^2 \, dY = \frac{1}{24}b^2 h^2$$

The above properties for the reference axes are easier to obtain by integration of the calculus definitions than are the properties for the centroidal axes by integration. The properties for the centroidal axes are needed, however, and they are most easily obtained by using the transfer axes formulas defined below.

Steel rolled sections(L and W sections, for example) are composed of two or more rectangles with some rounded corners(fillets) where the rectangles are joined together in the steel section. The properties of a rectangle and the fillets are obained by integration of the calculus definitions and the summation equation definitions given below are used to find the properties of a steel rolled section.

B.4 TRANSFER AXES FORMULAS

Definitions:

$$I_X = I_x + A\,\bar{Y}^2; \quad I_Y = I_y + A\,\bar{X}^2; \quad I_{XY} = I_{xy} + A\,\bar{X}\bar{Y}$$

NOTE: $\bar{X}$ and $\bar{Y}$ have signs. $\rightarrow\rightarrow\rightarrow\rightarrow\rightarrow\rightarrow\rightarrow\rightarrow\uparrow\uparrow$

$\bar{X}$ and $\bar{Y}$ can be measured from: X,Y origin to x,y origin

or from: x,y origin to X,Y origin.

However, both $\bar{X}$ and $\bar{Y}$ must be measured from same origin.

For Figure B1:

$$I_x = I_X - A\,\bar{Y}^2 = \frac{1}{12}bh^3 - \left(\frac{1}{2}bh\right)\left(\frac{h}{3}\right)^2 = \frac{1}{36}bh^3$$

$$I_y = I_Y - A\,\bar{X}^2 = \frac{1}{12}hb^3 - \left(\frac{1}{2}bh\right)\left(\frac{b}{3}\right)^2 = \frac{1}{36}hb^3$$

$$I_{xy} = I_{XY} - A\,\bar{X}\bar{Y} = \frac{1}{24}b^2h^2 - \left(\frac{1}{2}bh\right)\left(\frac{b}{3}\right)\left(\frac{h}{3}\right) = -\frac{1}{72}b^2h^2$$

It should be noted that the definitions assume the properties for the centroidal axes are known and the properties for the reference axes are to be obtained. That is, the transfer is made from the centroidal axes to the reference axes. Then, using algebra, the last three equations above were obtained from the definitions. Using the last three equations above, the properties for the centroidal axes are easily obtained from the properties for the reference axes.

B.5 SUMMATION FORMULAS

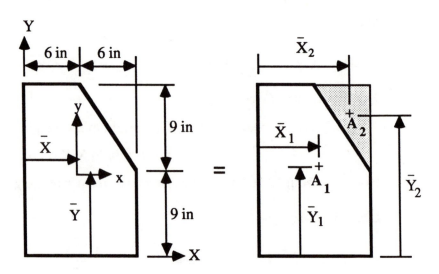

Figure B4-a   Figure B4-b

For Figure B4–b:

$A_1 = 12*18 = 216 \text{ in}^2;$ $\bar{X}_1 = 6 \text{ in.};$ $\bar{Y}_1 = 9 \text{ in.}$

$A_2 = -\dfrac{1}{2}*6*9 = -27 \text{ in}^2;$ $\bar{X}_2' = 10 \text{ in.};$ $\bar{Y}_2 = 15 \text{ in.}$

$$A = \sum_{i=1}^{2} A_i = 216 + (-27) = 189 \text{ in}^2$$

$$\bar{X} = \dfrac{\displaystyle\sum_{i=1}^{2} \bar{X}_i A_i}{A} = \dfrac{6*216 + 10*(-27)}{189} = 5.43 \text{ in.}$$

$$\bar{Y} = \dfrac{\displaystyle\sum_{i=1}^{2} \bar{Y}_i A_i}{A} = \dfrac{9*216 + 15*(-27)}{189} = 8.14 \text{ in.}$$

$$I_x = \sum_{i=1}^{2} (I_x + A \bar{y}^2)_i$$

$$I_x = \frac{1}{12}*12*(18)^3 + 216*(9 - 8.14)^2$$

$$- \left[\frac{1}{36}*6*(9)^3 + 27*(15 - 8.14)^2 \right] = 4599.6 \text{ in}^4$$

$$I_y = \sum_{i=1}^{2} (I_y + A \bar{x}^2)_i$$

$$I_y = \frac{1}{12}*18*(12)^3 + 216*(6 - 5.43)^2$$

$$- \left[\frac{1}{36}*9*(6)^3 + 27*(10 - 5.43)^2 \right] = 2044.3 \text{ in}^4$$

$$I_{xy} = \sum_{i=1}^{2} (I_{xy} + A \bar{x}\bar{y})_i$$

$$I_{xy} = 0 + 216*(6 - 5.43)*(9 - 8.14)$$

$$- \left[-\frac{(6)^2(9)^2}{72} + 27*(10 - 5.43)*(15 - 8.14) \right] = -700.1 \text{ in}^4$$

NOTE: I_{xy} for A_1 and A_2 were obtained from Figures B5 and B6.

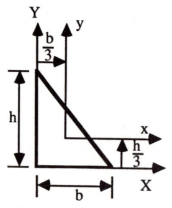

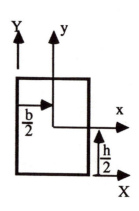

$$I_x = \frac{1}{36}bh^3; \qquad I_X = \frac{1}{12}bh^3$$

$$I_y = \frac{1}{36}hb^3; \qquad I_Y = \frac{1}{12}hb^3$$

$$I_{xy} = -\frac{1}{72}b^2h^2; \qquad I_{XY} = \frac{1}{24}b^2h^2$$

$$I_x = \frac{1}{12}bh^3; \qquad I_X = \frac{1}{3}bh^3$$

$$I_y = \frac{1}{12}hb^3; \qquad I_Y = \frac{1}{3}hb^3$$

$$I_{xy} = 0; \qquad I_{XY} = \frac{1}{4}b^2h^2$$

Figure B5_____

Figure B6_____

B.6 MORE ABOUT THE PRODUCT OF INERTIA

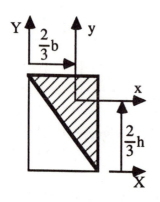

$$I_x = \frac{1}{36}bh^3$$

$$I_y = \frac{1}{36}hb^3$$

$$I_{xy} = -\frac{1}{72}b^2h^2$$

As shown below, there are two ways by which I_{xy} can be obtained.

Figure B7

1. Find I_{xy} for Figure B7 by using the definition:

302

$$I_{xy} = \sum_{i=1}^{2} (I_{xy} + A\,\overline{x}\,\overline{y})_i$$

$$I_{xy} = 0 + (by)*\left(\frac{2}{3}b - \frac{b}{2}\right)*\left(\frac{2}{3}h - \frac{h}{2}\right)$$

$$-\left[-\frac{1}{72}b^2h^2 + \frac{bh}{2}*\left(\frac{2}{3}b - \frac{b}{3}\right)*\left(\frac{2}{3}h - \frac{h}{3}\right)\right]$$

$$I_{xy} = -\frac{1}{72}b^2h^2$$

2. Find I_{xy} for Figure B7 by using mathematical logic:

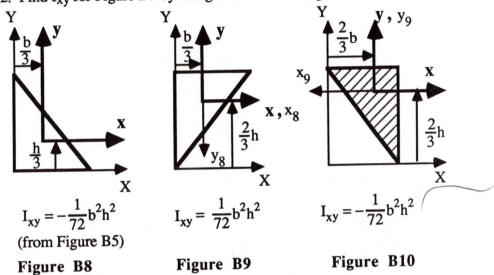

$$I_{xy} = -\frac{1}{72}b^2h^2$$
(from Figure B5)

Figure B8

$$I_{xy} = \frac{1}{72}b^2h^2$$

Figure B9

$$I_{xy} = -\frac{1}{72}b^2h^2$$

Figure B10

If the triangle and **x,y** axes in Figure B8 are rotated 180° about the **x** axis, we obtain the triangle and x_8, y_8 axes in Figure B9. Note that: in Figure B9, **x** and x_8 have the same positive direction, but **y** and y_8 are positive in opposite directions. Since one sign change occurred in obtaining Figure B9 by rotating Figure B8 about its **x** axis, change the sign of I_{xy} in Figure B8 to obtain I_{xy} for Figure B9.

If the triangle and **x,y** axes in Figure B9 are rotated 180° about the **y** axis, we obtain the triangle and x_9, y_9 axes in Figure B10. Note that: in Figure B10, **y** and y_9 have the same positive direction, but **x** and x_9 are positive in opposite directions. Since one sign change occurred in obtaining Figure B10 by rotating Figure B9 about its **y** axis, change the sign of I_{xy} in Figure B9 to obtain I_{xy} for Figure B10. Method 1 and Method 2 give the same sign for I_{xy} in Figure B10.

B.7   PRINCIPAL AXES

The *principal axes* are those <u>centroidal axes</u> for which the <u>product of inertia is zero</u>.
　　　Let: **u** denote the **major** principal **axis**(maximum inertia axis)
　　　　　　v denote the **minor** principal **axis**(minimum inertia axis)
Then, $I_{uv} = 0$. See Figure B11 for an example.

B.8   USING MOHR'S CIRCLE TO FIND THE
　　　　 PRINCIPAL AXES

EXAMPLE

　　　Given: The values for I_x, I_y, and I_{xy} shown in Figure B11.
　　　Find: Numerical values for α, I_u, and I_v using Mohr's Circle.

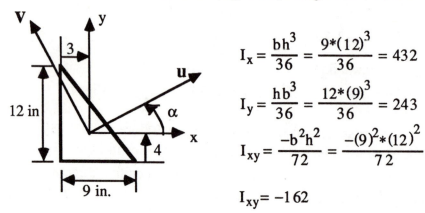

$$I_x = \frac{bh^3}{36} = \frac{9*(12)^3}{36} = 432$$

$$I_y = \frac{hb^3}{36} = \frac{12*(9)^3}{36} = 243$$

$$I_{xy} = \frac{-b^2h^2}{72} = \frac{-(9)^2*(12)^2}{72}$$

$$I_{xy} = -162$$

Figure   B11_____

304

SOLUTION -- The following described steps are illustrated in Figures B12 and B13.

Step 1 -- Plot point: $\left[(I_x, I_{xy}) = (432, -162)\right]$

Plot point: $\left[(I_y, -I_{xy}) = (243, 162)\right]$

Step 2 -- Locate center of circle:
$$C = \frac{I_x + I_y}{2} = \frac{432 + 243}{2} = 337.5$$

Step 3 -- Draw circle(see Figure B12): center is at C; circle passes through points plotted in Step 1

Product of Inertias

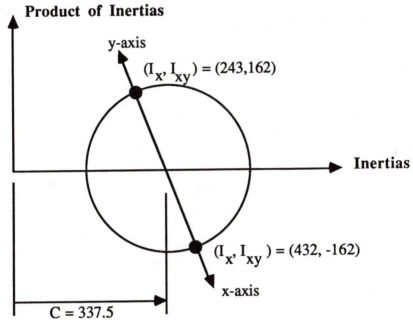

Figure B12

Step 4 -- See Figure B13. From point (I_x, I_{xy}), erect a perpendicular to the Inertias axis.

Calculate: $b = I_x - C = 432 - 337.5 = 94.5$
b is the base of the shaded triangle.
triangle height, **h** = (absolute value of I_{xy}).

305

$$R = \sqrt{b^2 + h^2} = \sqrt{(94.5)^2 + (162)^2} = 187.5$$

$$\tan 2\alpha = \frac{h}{b} = \frac{162}{94.5} = 1.714$$

$$2\alpha = 59.74° \text{ (on Mohr's Circle)}$$

$$\alpha = 29.87° \text{ (on Figure B11)}$$

Note:

Each angle in Fig B11 is **doubled** on Mohr's Circle.

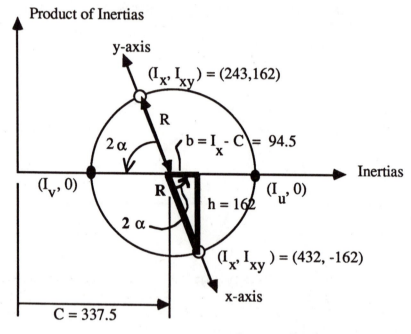

Product of Inertias

y-axis

$(I_{x'}, I_{xy}) = (243,162)$

R

$b = I_x - C = 94.5$

$(I_v, 0)$

R

$h = 162$

$(I_u, 0)$

Inertias

2α

2α

$(I_{x'}, I_{xy}) = (432, -162)$

x-axis

$C = 337.5$

$$\tan 2\alpha = h/b = 162/94.5 = 1.714$$

Figure B13

Step 5 -- In Figure B13, note that 2α on Mohr's Circle is measured CCW from the x-axis to the u-axis(maximum moment of inertia axis).

Also, as shown in Figure B13, 2α is measured CCW from the y-axis to the v-axis(minimum moment of inertia axis).

306

In Figure B11, at $\alpha = 29.87°$ CCW from the x-axis, draw the u-axis. Similarly, at $\alpha = 29.87°$ CCW from the y-axis, draw the v-axis.

Calculate: $I_u = C + R = 337.5 + 187.5 = 525$
$I_v = C - R = 337.5 - 187.5 = 150$

B.9 RADIUS OF GYRATION

If the cross section for which I_u and I_v was found by using Mohr's Circle is the cross section of a column, the radius of gyration is needed for each principal axis.

$$\textbf{Definition:} \quad r_u = \sqrt{\frac{I_u}{A}} \; ; \quad r_v = \sqrt{\frac{I_v}{A}}$$

where: $\mathbf{r_u}$ is the radius of gyration for the u-axis(**major principal axis** or maximum moment of inertia axis)

$\mathbf{r_v}$ is the radius of gyration for the v-axis(**minor principal axis** or minimum moment of inertia axis)

B.10 PROPERTIES OF A STEEL L SECTION

See the properties for an L9X4X5/8 shown on LRFD pages 1-50,51. Note that properties are given for the centroidal axes and for the **minor principal axis**(unfortunately named **Z** on LRFD page 1-51). Recall that in the author's notation, the **minor principal axis** is called **V** and the **major principal axis** is called **u**.

Using the definition of radius of gyration, we note that:

$r_v = r_z$ and $I_v = A r_v^2 = 7.73*(0.847)^2 = 5.55$

Then, using $\tan \alpha = 0.216$, $\alpha = 12.19°$, $2\alpha = 24.38°$, and Mohr's Circle(see Figure B14), we can find I_{xy} and I_u:

$$C = \frac{I_x + I_y}{2} = \frac{64.9 + 8.32}{2} = 36.6$$
$$R = C - I_v = 36.6 - 5.55 = 31.05$$
$$I_u = C + R = 36.6 + 31.05 = 67.7$$

On LRFD page 1-51, note that the Z-axis(minor principal axis which the author has renamed as **V**) is located at α = 12.19° <u>CCW from the y-axis</u>. The major principal axis, **u**, is not shown on LRFD page 1-51, but **u** is located at α = 12.19° <u>CCW from the x-axis</u>. On Mohr's Circle(see Figure B14), we need to locate the x-axis in order to find I_{xy} from the point whose coordinates are (I_x, I_{xy}). We know that the u-axis passes through the maximum moment of inertia point on Mohr's Circle and that the x-axis must be properly located such that the u-axis is at 2α = 24.38° <u>CCW from the x-axis</u>. Since h = R sin 2α = 12.82 and since the point (I_x, I_{xy}) lies in the region where the product of inertias is negative, then I_{xy} = -12.82. (Alternatively, I_{xy} can be found by using the summation formulas and the figure on LRFD page 1-51 decomposed into two rectangles, but this method takes more time than the method shown(using Mohr's Circle).

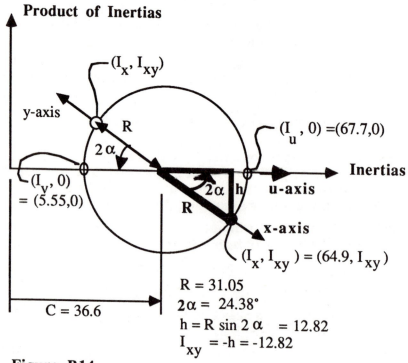

Product of Inertias

$-(I_x, I_{xy})$

y-axis

R

2α

$(I_u, 0) = (67.7, 0)$

Inertias

$(I_y, 0)$
$= (5.55, 0)$

2α   h

u-axis

R

x-axis

$(I_x, I_{xy}) = (64.9, I_{xy})$

R = 31.05

C = 36.6

2α = 24.38°

h = R sin 2 α = 12.82

I_{xy} = -h = -12.82

Figure B14

If the cross section of a member is a single angle, only the properties for the principal axes are needed and I_{xy} is not needed. However, if a single angle is fillet welded to a C section(see LRFD pages 1-106,110) or to some other rolled section to create a combined section, I_{xy} of the single angle will be needed if the properties for the principal axes of the combined section are needed.

B.11 FLEXURAL FORMULA

Consider a W18X50(see LRFD page 1-26) used as a simply supported beam(see LRFD page 3-130 Case 1). For discussion purposes, choose:

    L = 30 ft

    w = 1.5 k/ft(causes bending about only the x-axis in the figure on LRFD page 1-26 and in Figure B15)

Therefore, the maximum bending moment is:

$$M_x = \frac{wL^2}{8} = \frac{1.5*(30)^2}{8} = 168.75 \text{ ft k} = 2025 \text{ ink}$$

Note that the x and y axes in the figure on LRFD page 1-26 are axes of symmetry. Therefore, x and y are principal axes.

The flexural formula,

$$f = \frac{M*c}{I}$$

is only valid:

    1. in the linearly elastic region of the stress-strain curve

    2. for each principal axis

The parameters in the flexural formula are:

    f is the extreme fiber bending stress(either compression or tension)

    M is the bending moment about a principal axis

    c is the perpendicular distance from the bending axis to the extreme fiber for which **f** is being computed

    I is the moment of inertia for the bending axis

Suppose a section is bending about the <u>major principal axis</u>, **u** ; then the flexural formula is:

$$f = \frac{M_u * c_u}{I_u}$$

Note that M, c, and I are for the same principal axis, **u**. A similar formula can be written for the other principal axis, **v**, by substituting **v** for the subscript **u** in the preceding formula. The bending stress diagram(see Figure B15) varies linearly from zero at the bending axis to **f** at the extreme fiber. On one side of the bending axis, **f** is a compressive bending stress and on the other side of the bending axis **f** is a tensile bending stress. For a W section subjected to bending about the major principal axis, **x**, the LRFD Manual defines S_x as the elastic section modulus for x-axis bending:

$$S_x = \frac{I_x}{c_x} \quad \text{where } c_x = \frac{d}{2}$$
$$\text{then, } f = \frac{M_x * c_x}{I_x} = \frac{M_x}{S_x}$$

and S_x is given in the W section properties table. For a W18X50, see LRFD page 1-27, $S_x = 88.9 \text{ in}^3$. For $M_x = 168.75$ ftk = 2025 ink,

$$f = \frac{2025 \text{ ink}}{88.9 \text{ in}^3} = 22.8 \text{ ksi}$$

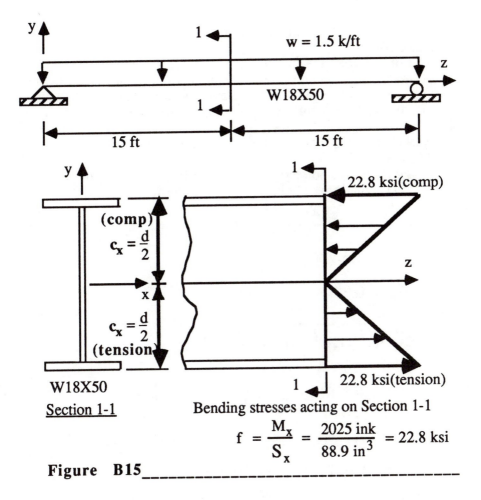

$$f = \frac{M_x}{S_x} = \frac{2025 \text{ ink}}{88.9 \text{ in}^3} = 22.8 \text{ ksi}$$

Figure B15_____

B.12 BIAXIAL BENDING

Since the author shows only one example of biaxial bending of a beam, the example entails a completely general case of a single angle, L9X4X5/8, loaded through the shear center to prevent twisting of the cross section and to avoid the need for computing torsional stresses(see Figure B16). The most convenient manner to fillet weld a single angle to another steel section having flat exterior surfaces is to lay the longer leg of the single angle on a flat region of the other steel section and fillet weld at each end of the longer leg of the single angle. An example of biaxial bending of a single angle used as a compression member in a

311

truss is given on LRFD page 2-49. Such a truss member is classified as a beam-column.

The objective of the following example is to illustrate how to use the flexural formula to compute the total bending stress at a point on a cross section which is subjected to biaxial bending. Since the flexural formula is only valid for each principal axis(see Section B.11), we need:

$I_u = 67.7$ and $I_v = 5.55$ found in Section B.10

M_u and M_v

c_u and c_v for the point where the total stress is to be computed
    (the chosen point is at the heel corner in Figure B16)

$$M_x = \frac{wL^2}{8} = \frac{1*10^2}{8} = 12.5 \text{ ftk} = 150 \text{ ink}$$

$\alpha = 12.19°$

$M_u = M_x*\cos \alpha = 146.6 \text{ ink}; \quad M_v = M_x*\sin \alpha = 31.7 \text{ ink}$

$$D = \sqrt{(0.858)^2 + (3.36)^2} = 3.47$$

$$\tan \theta = \frac{0.858}{3.36} = 0.255; \quad \theta = 14.32°$$

$\phi = \theta + \alpha = 26.5°$

$c_u = D*\cos \phi = 3.11; \quad c_v = D*\sin \phi = 1.55$

At the heel corner of the L9X4X5/8,

   M_u produces a tensile stress,

   and M_v produces a tensile stress.

$$f = \frac{M_u c_u}{I_u} + \frac{M_v c_v}{I_v}$$

$$f = \frac{146.6*3.11}{67.7} + \frac{31.7*1.55}{5.55}$$

$f = 6.73 + 8.85 = 15.6 \text{ ksi(tension)}$

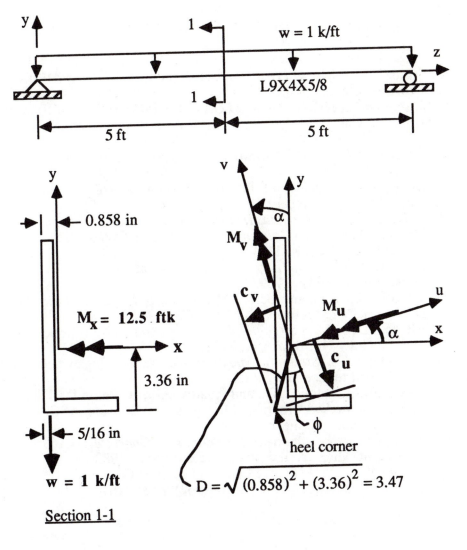

$$D = \sqrt{(0.858)^2 + (3.36)^2} = 3.47$$

Section 1-1

Figure B16_____

References

1. Smith, J.C., *Structural Analysis* , New York,NY: Harper & Row Publishers, Inc., 1988.
2. *Manual of Steel Construction, LOAD & RESISTANCE FACTOR DESIGN*, Chicago, IL:American Institute of Steel Construction, 1986.
3. Galambos, T.V., and Ellingwood, B., "Serviceability Limit States:
   Deflection," *Journal of the Structural Division*, ASCE,
   Vol. 112, No. 1, January 1986, pp. 68-84.
4. Euler, Leonhard, *Methodus Inveniendi Lineas Curvas Maximi Minimive Proprietate Gaudentes*, Appendix: *De Curvis Elasticis*, Lausanne and Geneva, 1744.
5. Euler, Leonhard, "Sur la force de colonnes," *Memoires de l 'Academie de Berlin*, 1759.
6. Timoshenko, S.P., and Gere, J.M., *Theory of Elastic Stability*, New York, NY: McGraw-Hill Book Co., 1961.
7. Bleich, F., *Buckling Strength of Metal Structures*, New York, NY: McGraw-Hill Book Co., 1952.
8. Galambos, T.V., *Structural Members and Frames*, Englewood Cliffs, NJ: Prentice-Hall, 1968.
9. Wood, B.R., Beaulieu, D., and Adams, P.F., "Column Design by P-Delta Method," Journal of the Structural Division, ASCE, Vol. 102, No. ST2, February 1976.
10. *Building Code Requirements for Reinforced Concrete*, ACI 318-83, Detroit, Mich.: American Concrete Institute, 1983
11. Kavanagh, T.C., "Effective Length of Framed Columns", *Transactions*, ASCE, 127, Part II(1962), pp81-101.

Index

319

WIDENER UNIVERSITY
WOLFGRAM
LIBRARY
CHESTER, PA.